S0-AWV-368

Surface Engineering
Volume II: Engineering Applications

Surface Engineering
Volume II: Engineering Applications

Edited by

P.K. Datta and J.S. Gray

University of Northumbria at Newcastle, Newcastle upon Tyne

Based on the Proceedings of the Third International Conference on Advances in Coatings and Surface Engineering for Corrosion and Wear Resistance, and the First European Workshop on Surface Engineering Technologies and Applications for SMEs. Both events were held at the University of Northumbria at Newcastle, Newcastle upon Tyne, UK, on 11–15 May 1992.

The front cover illustration was kindly provided by Tecvac Ltd., UK.

Special Publication No. 127

ISBN 0-85186-675-1

A catalogue record for this book is available from the British Library

Published by The Royal Society of Chemistry,
Thomas Graham House, Science Park, Cambridge
CB4 4WF

Printed in Great Britain by Redwood Books Ltd., Trowbridge, Wiltshire

Preface

Surface Engineering is an enabling technology widely applicable to a range of industrial sector activities. It encompasses techniques and processes capable of creating and/or modifying surfaces to provide enhanced performance such as wear, corrosion and fatigue resistance and bio-compatibility. Surface engineering processes can now be used to produce multilayer and multicomponent surfaces, graded surfaces with novel properties and surfaces with highly non-equilibrium structures. In a broad sense surface engineering covers three interrelated activities:-

1. *Optimization of surface/substrate properties and performance* in terms of corrosion, adhesion, wear and other physical and mechanical properties.

2. *Coatings technology* including the more traditional techniques of painting, electroplating, weld surfacing, plasma and hypervelocity spraying, various thermal and thermochemical treatments such as nitriding and carburizing, as well as the newer processes of laser surfacing, physical and chemical vapour deposition, ion implantation and ion mixing.

3. *Characterization and evaluation of surfaces and interfaces* in terms of composition and morphology, and mechanical, electrical and optical properties.

It is now widely recognized that the successful exploitation of these processes and coatings may enable the use of simpler, cheaper and more easily available substrate or base materials, with substantial reduction in costs, minimization of demands for strategic materials and improvement in fabricability and performance. In demanding situations where the technology becomes constrained by surface-related requirements, the use of specially developed coating systems may represent the only real possibility for exploitation.

The three volumes of ***"Surface Engineering"*** have been prepared focusing attention on these comments and have been principally based on papers presented at the 3rd International Conference on Advances in Coatings and Surface Engineering for Corrosion and Wear Resistance. The structure and contents of the volumes in this series have been conceived to provide a number of interrelated themes and a coherent philosophy. Additional material has been incorporated to complement the information delivered at the Conference. As such, the text provides a useful blend of Keynote, review, scientific and state-of-the-art type papers by international authorities and experts in surface engineering.

"Surface Engineering" is structured in three volumes. The first volume, *Fundamentals of Coatings* considers principles of coating/substrate design in high temperature and aqueous corrosion and wear environments, and scans the coatings' spectrum from organic, through metallic to inorganic. Here there is a general emphasis on the science and design of coating/substrate systems rather than technology. The second volume, *Engineering Applications*, is dedicated to topics concerning the performance of coatings and surface treatments embracing four main areas - the inhibition of wear and fatigue, corrosion control, application of coatings in heat engines and machining, and quality and properties of coatings. Finally, the third volume has two main thrusts: *Process Technology and Surface Analysis*. Both areas are clearly central to surface engineering, and each holds particular promise, not only for improvements in existing types of coatings' performance, but also in the design, development and evaluation of totally new - for example hybrid - coating/substrate systems.

The editors wish to pay tribute to Dr. Tom Rhys-Jones who recently died.

Contents

PART 2: ENGINEERING APPLICATIONS

Section 2.1 Wear and Fatigue Resistance

Section 2.2 Corrosion Control

Section 2.3 Heat Engines

Section 2.4 Machining

Contents
Volume I: Fundamentals of Coatings

PART 1: FUNDAMENTALS OF SURFACE COATINGS

Introduction

Section 1.1 High Temperature Corrosion

Section 1.2 Aqueous Corrosion

Section 1.5 Metallic Coatings

Section 1.6 Ceramic and Glass Ceramic Coatings

Contents
Volume III: Process Technology and Surface Analysis

Section 4.3 Surface Integrity and Properties

Acknowledgements

The editors wish to express their gratitude for the support extended by:

> The Commission of the European Communities, The Department of Trade and Industry, The Institute of Materials, The Institute of Corrosion, Northern Electric, METCO Ltd., Cobalt Development Institute, LECO Instruments and Severn Furnaces.

Special thanks are due to Dr N E W Hartley of the CEC for his continual work and support of surface engineering.

The support and encouragement of Dr C Armstrong, Head of the Department of Mechanical Engineering and Manufacturing Systems, is gratefully acknowledged. Thanks are also due to Prof J Rear for opening the Conference.

The human commitment to any conference or book is substantial and often not fully acknowledged. In this final regard the work of Kath Hynes, Pauline Bailey, the technicians from the Department of Mechanical Engineering and Manufacturing Systems and the members of the Surface Engineering Research Group should be fully recognized.

Finally special commendation is reserved for Lee Comstock who administered the 3rd International Conference on Advances in Coating and Surface Engineering for Corrosion and Wear Resistance and the 1st European Workshop on the Application of Surface Engineering for SMEs, and was also responsible for the retyping of many conference papers.

P.K. Datta
J.S. Gray
University of Northumbria at Newcastle

Part 2: Engineering Applications

Section 2.1 Wear and Fatigue Resistance

2.1.1
Production of Corrosion and Wear Resistant Coatings

M.G. Hocking

MATERIALS DEPARTMENT, IMPERIAL COLLEGE, LONDON SW7 2BP, UK

1 INTRODUCTION

Materials are approaching their upper performance limits in all fields. Further improvements require the development of composites, of which coated materials are one form. The coating of components to protect them from corrosion and abrasion is increasing. This has considerable economic importance.

The first comprehensive review on metallic and ceramic coatings, containing over 2000 references, has recently been published as a book[1] and references to all the topics described in this paper can be found in this book, which includes a comprehensive set of figures.

Coatings are used both in high temperature and aqueous systems. Electric power generation by coal gasification involves severe conditions and thick coatings have proved effective. Gas turbine and diesel engines are subject to high temperature corrosion and successful coatings have proved very beneficial. Nuclear power systems also rely on coatings. Problems include: porosity, adhesion, substrate compatibility, cost, possibility of renewal or repair, interdiffusion, effect of thermal cycling and corrosion.

2 METALLIC COATING SYSTEMS

Diffusion Coating Systems

Aluminizing forms an alumina scale-producing surface which is the best protection for high velocity gas turbine blades, but a diffusion barrier coating is needed. Al coating on steel is good to 500^{0}C above which brittle intermetallics are formed. Chromised steel (diffused in) is good up to 700^{0}C and can be bent 180^0 without damage but above 800^{0}C Cr diffuses further into the steel and reduces oxidation resistance and forms brittle intermetallics. Adding Al or Si offsets this effect.

On aluminized superalloys, heat treatment is given to stabilise the NiAl phase. Adding Cr and Ti improves the hot corrosion resistance and dispersed oxide particles (e.g. Y_2O_3) reduce spalling. Aluminide coatings on superalloys lack ductility and spall below 750^{0}C; to stop this the brittle beta NiAl or CoAl are embedded in a ductile gamma solid solution matrix. Adding Y improves adherence.

Diffusion coatings used successfully on early gas turbine blades were tied to the substrate composition and microstructure. Later, the substrate alloys had less Cr and more refractory metals and castings had more segregation. This required coatings which were much more independent of substrate. Overlay coatings met this necessity.

Overlay Coatings

MCrAlY (M = Ni, Co, Fe) alone or in combination coatings by electron beam evaporation have successfully coated over 3 million aerofoils (1982 figure). They are multiphase ductile matrix (e.g. gamma CoCr) alloys containing a high fraction of brittle phase (e.g. beta CoAl). Their success is in the presence of Y and Hf which promote alumina adherence during thermal cycling.

Overlay coatings are diffusion bonded to the substrate but not with the intention to convert the whole coating thickness to NiAl or CoAl. So there is more freedom in coating composition.

Plasma Sprayed Coatings have considerable porosity which lowers strength and decreases corrosion resistance.

Ion Implantation of Al in Fe and of Y in NiCr, FeNiCr, etc reduces oxidation and Y improves scale adherence.

Surface Welded or Clad systems are good for wear and corrosion resistance, usually > 3 mm thick. Multilayers of hard alloys interleaved with ductile layers stop cracks.

3 CERAMIC COATINGS

Wear resistant coatings are economically important - in USA machining costs are $70 billion/year (1978) and tools cost $1 billion/year (half each: cemented carbide and tool steel).

Vehicle engines need ceramic thermal barrier protection of piston caps, cylinder shields and liners etc.

TiC and TiN on steel have been coated by ARE, RS and CVD (see below). TiC coated stainless steel wears 34 times better than hard Cr plate.

Ceramic coatings tend to be thick and brittle and to devitrify and to spall on thermal cycling, but they offer excellent resistance to oxidation and corrosion and are very suitable for static components like combustion liners. Further research is needed for rotating components.

Ceramic coatings are often complex silicates plus zirconia, titania, alumina, ceria, etc and protect up to 1260^0C. Nitrides give wear resistance and a thermal barrier. Elements are added to improve crack tolerance.

Vitreous enamel on steel is widely used, up to 600^0C. Silicides and borides are also commonly used. Siliconizing is also good for refractory metals as $MoSi_2$, WSi_2 and VSi_2 have good oxidation resistance.

Rubbing seals are used in ceramic regenerators in gas turbines and rubbing surfaces are needed in many other cases.

Present thermal barrier coatings for gas turbines are a plasma sprayed zirconia - 7% yttria layer over a MCrAlY (M = Ni, Co or NiCo) bond-coat layer plasma sprayed at low pressure. Adhesion of plasma sprayed ceramic coatings to metals is generally poor but is improved by a sprayed bond-coat or interlayer (e.g. NiAl or Mo).

Diffusion Barrier Coatings

TiC is a carbon diffusion barrier for Ta coatings on steel and TiN is a nitrogen barrier on SiC heating elements.

Desirable Ceramic Coatings Properties

Elasticity is desirable to prevent thermal cycle spalling. Elasticity of ceramics rises with porosity but open porosity invites corrosion. Self-sealing ceramic coatings give a solution due to closed porosity. A graded bond to the metal is essential.

4 COATING PROCESSES

Physical Vapour Deposition (PVD)

PVD covers 3 major techniques: evaporation, sputtering and ion plating. The steps are:

1. Synthesis of the material to be deposited - conversion to vapour phase and if a compound is to be deposited this way includes a vapour phase reaction to form it.
2. Transport of vapour to substrate.
3. Condensation.

These steps can be controlled independently in PVD processes, which is an advantage over CVD. Conventional PVD is an overlay coating. Purity, structure and adhesion can all be controlled.

Vacuum evaporation is much faster than vacuum sputtering but coating composition and deposition rate are easier to control by sputtering. Ion plating gives best adhesion due to its high deposition energy.

PVD causes lower substrate temperature (e.g. 500^{0}C) than CVD (e.g. 1000^{0}C). 500^{0}C is lower than the steel tempering temperature but for CVD that is exceeded and a quench is needed which may cause distortion and adhesion loss.

The advantages of PVD over other overlay coatings processes like electrodeposition, CVD and plasma spray are:

1. Extreme versatility in deposit composition. Almost any alloy, refractory and some polymers can easily be deposited. In this, PVD is better than any other coating process.
2. Substrate temperature can be varied over wide limits.
3. Vapoforms can be produced at high rates.
4. Very high purity of the deposits.
5. Excellent bonding to substrate is possible.
6. Excellent surface finish.

Evaporation

The working pressure is 1 mtorr to 1 torr. Nucleation and growth kinetics govern the evolution and microstructure of the coating. Evaporation is by resistance heating from a foil or boat of W, Mo, Ta, or if these react with the liquid pool a refractory boat can be used (e.g. BN-TiB_2 which is electrically conducting is used for Al). Other heating methods include electron beam heating by thermionic gun or plasma gun:

Thermionic Gun. Electrons emitted from a hot cathode are attracted towards an anode at 10 to 40 kV at 10^{-5} torr (negligible ionisation). The anode is the evaporant.

Plasma Electron Beam Gun. An electric field applied to a low pressure gas generates electrons. The large number of ions also present cause high coating adhesion. Many designs give improved performance. Electron Beam PVD CoCrAlYTa and NiCrAlYSi are more protective than the plasma sprayed versions but a high vacuum is necessary, unlike for sputtering. Twenty five microns/minute is possible but limited to line-of-sight. Millions of aero gas turbine blades have been coated.

Reactive Evaporation. A metal or its compound is evaporated in the presence of some reactive gas to form a compound - e.g. Si or SiO in oxygen to form silica coating, Ti in nitrogen to form TiN_2 coating. The mean-free-path exceeds the source-to-substrate distance and so reaction only occurs on the substrate (unlike CVD).

Activated Reactive Evaporation (ARE) is similar but at a slightly higher pressure with mean-free-path less than source-to-substrate distance so that reaction occurs in the gas phase before deposition (as with CVD). This allows stoichiometry variation, e.g. of TiC coatings.

Biased Activated Reactive Evaporation or Reactive Ion Plating. The substrate is negatively biased to attract the positive ions in the plasma. Other variants are also in use.

Sputtering

This is a momentum transfer process in which a fast ion (e.g. Ar^+) hits an atom from a cathodic surface and sputters it out towards a substrate. Sputtered films have compressive stress while evaporated films have a tensile stress. Low stress is good for adhesion but compressive stress may reduce cracking. Carbon with a diamond structure has been sputtered onto steel.

Magnetron Sputtering creates fields which cause the electrons (emitted from the cathode) to follow long helical paths which give many more Ar^+ ions for sputtering. Scale-up allows inside coating of 2 m long steel tubes.

Many improved designs have been made. The substrate is cleaned by ionic bombardment prior to the sputtering. Adherent dense coatings are obtainable at lower substrate temperatures than by CVD. Type 304 stainless steel has been sputtered onto float glass substrate at room temperature.

Magnetron sputtered Cr in Ar (2 mtorr) + CH_4 or N_2 has hardness up to 3500 HV, with fretting, wear, fatigue and corrosion resistance better than electrolytic hard chrome.

Highly corrosion resistant films of amorphous $Co_{100-x}Nb_x$, $Fe_{80-x}Cr_xP_{13}C_7$, $Cr_{75}B_{25}$, $Ti_{25}B_{25}$alloys were sputtered onto water cooled Cu substrates.

RF Sputtering can be used for insulators (which cannot be d.c. sputtered because they "charge-up"). Electrons in the plasma prevent charging and ion plating then occurs in the appropriate half-cycles.

Reactive Sputtering (RS) occurs when one of the species is gaseous, e.g. sputtering Al in oxygen to form alumina coatings, Ti or Nb in nitrogen to form TiN or NbN.

Ion Beam Sputtering (Sputter Ion Plating) allows independent control of energy and current density of the bombarding ions. A vacuum chamber (1 Pa) at 300^0C is lined with coating source plates at -1kV (which causes ion bombardment and consequent sputtering of material from them). The coating substrates are held at -100V (to ion polish them without significant re-sputtering). No manipulation of the specimens is needed - there is good throwing power.

Cd electroplating can be advantageously replaced by ion vapour deposited Al (Cd on strong Al and steel alloys promotes exfoliation corrosion). Al coatings are usable up to 500^0C. Coating structure and stress are affected by substrate bias.

Ion plating with Al to protect U from oxidation has been successful in contrast to ordinary vacuum evaporation, due to a thin UO_2 film which forms immediately on U on cleaning. In Ar glow discharge ion plating, the specimen (U) surface is sputter cleaned by Ar^+ ions bombardment and is then held contamination-free until and during the ion plating process. The high energy Ar^+ collision with the surface heats it and promotes diffusion. The maximum substrate diameter is 25 cm at present. Ion plating is good for wear and erosion resistant coatings due to its good adhesion (e.g. HfN for low friction and TiC for deformation resistance and good bonding). For an incompatible substrate and coating, ion plating can be used as a strike for electroplating.

Advantages of ion plating:
- Good adhesion at low substrate temperatures.
- Good coverage of complicated shapes.
- Good structures and structure-dependent properties, at low substrate temperatures.

Disadvantages of ion plating:
- Must fix specimen to a high voltage electrode.
- Must work in relatively high gas pressure.
- Masking difficult.

- Deposition energies are in a wide range and are hard to control.

Ion Implantation

Chosen atomic species are ionised and then accelerated in a field of 10 to 1000 keV in a moderate vacuum (1mPa). The ion source and extraction electrodes are designed so that ions emerge as a beam. Ion penetration is only about 0.2 microns but materials properties are greatly altered.

Advantages:

- High temperatures not required, so no thermal distortions of substrate occur.
- Absence of any interface eliminates coatings adhesion problems.
- Surface finish is improved as sputtering polishes it.
- Implanted species is finely dispersed, which improves its effect during corrosion.
- Biaxial compressive stress is induced in the surface, which closes surface microcracks in ceramics. In metals these stresses can be annealed out.

Chemical Vapour Deposition (CVD)

This is a process where a solid reaction product nucleates and grows on a substrate, due to a gas phase reaction. Heat, plasma or ultraviolet light can enable it. Like electroforms, thick free-standing vapoforms can be produced, e.g. of W and W-Re in complex shapes. Many factors affect a good CVD coating, such as gas reactant composition, flow rate, Reynolds number, substrate temperature and total gas pressure. In consequence, many manufacturers keep the optimum parameters secret.

Typical CVD coatings include: B,C,Nb, Si, Ta, W, AlN, BP, BN, IrO_2, RuO_2, SiC, Si_3N_4, SiO_2, TiB_2, TiC, TiN, TiCN, ZrN, $ZrSi_2$, ZrC, B_4C, Cr carbides, $(Fe,Mn)_3C$, HfC, Mo_2C, NbC, SiC, TaC, TiC, VC, W_2C, ZrC, CrN, Fe_4N, HfN, NbN, Ta_xN, alumina, chromia, hafnia, niobia, titania, yttria, Fe_2O_3, In_2O_3, SnO_2, Ta_2O_5, V_2O_3, MoB, NbB_2, TaB_2, TiB_2, WB.

Substrates include Ti, Ta, Mo, W, C, Pt, Fe, Ni, Co, Mo, superalloys, Inconel, Mo, SiO_2, SiC, ceramics, glasses, BN, cemented carbides, Cu, Si, Nb, etc.

Possible bad effects on a substrate, of the high temperature of CVD, may be avoidable by applying plasma technology. Laser assisted CVD and metal organic CVD are also important methods. Some substrate/deposit reaction is expected, which may be beneficial for bonding. Usually a CVD reactor has a cold wall and only the substrate is heated so that deposition does not occur on the walls. Substrate preparation, heating and positioning are important factors. Turbulent gas flow is to be avoided for several reasons including its uncontrollable local cooling effects. The thermodynamic theory for CVD is well understood but the kinetics involves many factors and the overall process is difficult to predict.

Some advantages of CVD are:
- It is versatile and gives some coatings not easily obtained otherwise.
- CVD can be done on powders (fluidised bed reactors), wires, wafers, complicated shapes.
- CVD is not restricted to line-of-sight.
- Relatively simple equipment.
- Deposition is possible over a wide range of pressures and temperatures.
- There is a choice of chemical systems/reactions, so that (e.g.) H embrittlement can be avoided.
- CVD has good throwing power.
- Composition can be controlled to give graded coatings.
- Coating structure and grain size can be controlled.
- Uniform coating on a complex shape is possible.

Pack Coatings

There are several variants: vacuum pack, slip pack, pressure pulse pack, etc. The pack can be all solid, or part fluid (slip pack). It is a special type of CVD where the substrate is buried in a mass of the depositing medium, which is a mixture of the master alloy with a salt activator. This pack is then heated and pressures can be 1 torr to 1 bar. It is a convenient process for gas turbine blades and rocket/spacecraft hardware. Large items and intricate shapes can be coated. Commonest coatings are Al and Cr. During pack aluminizing of Ni superalloy turbine blades, Ni_2Al_3 and other intermetallic phases are formed. Good protection against superalloy hot corrosion is given by a duplex Si slurry and aluminide formed by a slurry or a spray of pure Si followed by a 16 hour pack aluminizing at 1100^{0}C. Steel can be pack boronized for wear resistance, giving well-bonded layers of FeB + FeB_2.

Pt-Al duplex coatings can be pack coated onto Ni-base superalloys to give improved DBTT. Pack chromizing of steel uses typically 50% Cr powder, 1% NH_4Cl and 49% alumina (inert filler to stop too fast coating). This gives 30 to 45% Cr in the surface.

Vacuum Pack Process needs no inert filler.

Pressure Pulse Pack evacuates and refills the system about 8 times/minute at 900^{0}C. It is especially suitable for coating crevices such as the cooling channels of gas turbine blades.

Slurry Coating uses a finer mesh metal mixture than pack coating and is at a temperature above the m.p. of this mixture.

Reaction Sinter Coating is an extension of slurry coating, designed to prevent the total loss of deposit identity which occurs in the diffusion process.

Sol-Gel Coating

This is probably the most economic process for high temperature ceramic coatings. A review has recently been published[2]. The method is in 4 stages: (i) sol preparation, (ii) spraying sol onto substrate, (iii) drying the sol to a gel, (iv) firing to obtain a coating.

Colloidal dispersion preparations are well characterised and can be aqua- or organo-sols, aggregate or non-aggregate in nature. They are stable dispersions of particles in a fluid of colloidal units of hydrous oxides or hydroxides with 20 Å to 1 micron particles. Sol preparation is either by hydrolysis followed by polymerisation, or by precipitation and then peptization. Only oxides have been sol-gel coated so far, but nitrides may be possible.

Sols are applied by immersion, spinning, electrophoresis or spraying. The sol-to-gel transformation is done by slow drying and then firing at 500 to 1200^{0}C for 15 minutes. A single such application gives 0.1 to 2 micron thick coating. Unlike slurry coats, there may not be a distinct diffusion zone, but the coating oxide and a growing oxide become interspersed. Sol-gel coatings have been characterised for oxidation resistance.

Hot Dip Coatings

These are traditionally applied to ferrous alloys. A metal such as Zn, Al, Sn or Pb is vat melted and the substrate is passed through. Substrate pre-cleaning is done by hot reducing gas (N_2 + H_2 at 800^{0}C) as a last stage following the conventional steel cleaning steps. The thermal cycling during 1 hot dipping of steel makes it less ductile but stronger. Brittle Fe_3Al forms and should be less than 10 microns thick if ductility is required. Hot dipping of a thermal barrier coating is done by dipping in a Zr-25wt% Ni eutectic with 5 wt% Y as stabilizer for the zirconia to be formed in the final conversion step. Superalloys were dipped at 1030^{0}C for 1 minute in Ar, then annealed for 4 hours and finally held in pO_2 of 10^{-17} bar ($H_2O/H_2/Ar$ mixture) to convert selectively to zirconia (without oxidising the Ni or Co), giving adherent and thermally cycle resistant coatings.

Electroplating and Chemical Coating

These will only be treated briefly as many reviews are available. Surface preparation before coating is essential. Etching up to 0.1 mm is necessary to get adhesion. The electrolyte condition is an important parameter and has been well characterised in standard texts. Composites such as cements can be electrodeposited by loading refractory powders or fibres into the electrolyte and keeping it well stirred. Electrodeposition of overlay coatings is a simpler process than PVD and CVD but because the deposition is a low energy (ambient temperature) process, adhesion may not be so good unless suitable heat treatment is carried out. Barrel plating for co-depositing 10 micron MCrAlY powder in a Co or Ni matrix has been used for gas turbine blades.

Electrophoretic composite coatings require fine particles in a liquid dielectric which migrate in an electric field and deposit on an electrode. High rates (1 micron/s) are possible, with good thickness control and throwing power but heat treatment is essential to sinter and densify.

The main drawback to electroplating is to get adequate throwing power. Interlayers may be needed; factors are:
- substrate/coating bonding, e.g. for substrates of stainless steel, Al-, W-, Nb-, Mo- alloys, etc; a copper strike suitable for low temperature use will be unsuitable at high temperature;
- possibility of substrate oxidation in the aqueous electrolyte;
- in-situ incorporation of S, P, H, or other undesirable radicals in the electroplate.

Electroless coating (or chemical coating) is a cold coating technique, e.g. of Ni, and has good throwing power giving very uniform coatings (unlike electroplating) on complex shapes and large components. It also gives a quasi-crystalline or near-amorphous deposit. Electroless Ni-P contains 7 to 10% P, but has good corrosion and wear resistance and low porosity.

Metalliding or Fused-salt Electroplating

Fe, Ni, Co, Mo, W, Nb, Ti, Cr, etc can be diffusion coated with Be, B, Al, Si, Ti, V, Cr, Mn, Y, Zr and rare earths, in fused fluorides or chlorides. Metalliding is of interest for surface hardening and for corrosion protection, e.g. Mo burns away in air at 650^{0}C but withstands 1300^{0}C for several hours if it is silicided.

Laser Coatings

A pure Fe surface can be precoated with DAG graphite and then laser surface melted (say) 12 times, re-applying graphite between each melting. Up to 6% C can thus be alloyed in (carburization).

Silica coatings have been laser fused onto Incolloy 800H. Dense zirconia layers have been deposited by pulsed laser. Laser cladding of triballoy, Ni alloys, WC + Fe, TiC, Si and alumina can be done by flame spraying followed by laser surface melting. Plasma sprayed porous coatings can be laser surface melted (without melting down to the interface) to seal surface porosity without removing bulk porosity (which confers some elasticity to ceramics). Mild steel has been laser clad with stainless steel. Powdered coating material can be blown onto a substrate by Ar, while applying a laser beam. The beam melts the powder which then falls onto the surface, protected by an Ar envelope.

Rapid Solidification Processing is achieved by a very large temperature difference between substrate and coating (originally in powder form, or from another type of coating method like PCV). The coating is thus made amorphous[1].

Metal Spraying

Plasma Spraying. Powder is carried in an inert gas stream into an atmospheric pressure of low pressure plasma. Any coating thickness is feasible. Problems include porosity and poor adhesion, especially of ceramics to metals. Adhesion is improved if a bond-coat interlayer is used (e.g. Ni-Al, Mo) and/or if small particle size powder is used. A plasma torch has a water cooled Cu anode and W cathode. Ar + H_2 flow round the cathode at 3.5 litres/minute and exit through the anode nozzle. A d.c. arc is maintained and the plasma of ions and atoms emerges at 6,000 to 12,000^{0}C at 1 cm from the nozzle, falling rapidly to 3,000^{0}C at 10 cm. Near the nozzle the gas velocity is 200 to 600 m/s and particle velocity is only 20 m/s, but the particle acceleration (due to gas frictional forces) is 100,000g so that 18 micron particles reach 275 m/s (maximum) at 6 cm from the nozzle. The very high temperature melts most refractory particles but low m.p. materials like plastics can also be sprayed. The plasma jet is 5 cm long and substrate distance is 15 cm. About 80 kW (typically 700 amps) is needed! Oxidation is minimised by the low dwell time and reducing gas. The substrate rarely exceeds 320^{0}C and is often $<$ 150^{0}C. Plasma spraying is the only spraying-type process which causes no distortion of the substrate.

For strong dense coatings, most of the particles must be molten before impingement and must have enough velocity to splat into the irregularities of the previous splats. Many coatings contain easily oxidised elements (e.g. Al, Mo, Ti) and then plasma spraying at low pressure (20 mbar) or in Ar is advantageous. Underwater plasma spraying offers much scope for deposition of wear and corrosion resistant coatings on submerged substrates such as offshore structures, e.g. with powders of self-fluxing Ni-based hardfacing alloys, Colomonoy 5,52 and 42 with metalloids B and Si to reduce oxide films on base metal. Coatings of 100 to 300 microns have been deposited.

Low Pressure and Vacuum Plasma Spraying are used for metals too reactive to be sprayed in air. The higher particle velocity gives high quality pure coatings. The jet velocity is mach 3, giving low porosity and good adhesion. The equipment is expensive.

D-gun Coating explosively propels molten powder particles onto the substrate. A water cooled gun barrel 1 m long, is fed with powder charge in acetylene + oxygen + nitrogen and this is repetitively exploded with a spark plug. This accelerates the powder to 720 m/s and melts it, so that it rapidly splats onto the substrate.

Advantages are:
- relatively simple equipment;
- low porosity, high bond strength coatings are produced;
- impurity tolerant;
- substrate pretreatment not stringent;
- strong bond with only moderate substrate heating during deposition;
- high coating rate.

All plasma and D-gun coatings have some open porosity which allows corrosion, especially by a molten salt. Porosity is reduced by sintering and other methods (see later).

Flame Spraying has 25 times less kinetic energy than D-gun, but is adequate for some purposes, e.g. for thermal barrier coatings on rocket nozzles and jet engine combustion chambers. A wire or powder feed of the coating material is fed into a flame and its end melts and is blown off onto the substrate by a fast compressed air blast. Uses include corrosion protection at ambient and high temperature, rebuilding worn parts and giving wear, erosion and abrasion resistance. Coatings range from Al to alumina. Porosity far exceeds that of D-gun coatings. The spray contains many non-molten particles, depending on the m.p. of the feed. Porosity can be reduced subsequently by resin vacuum impregnating; laser surface melting is also suitable.

Electric Arc Spraying. Two feedable wires are made electrodes for a d.c. arc and the resulting molten droplets are blown onto the substrate by a compressed air blast. The molten drops are coarser but hotter than those from flame spraying and adhesion is better. Pseudo alloy coatings are obtained by feeding different wires through each electrode. Thus a 2-phase alloy of Cu and stainless steel is produced, giving high wear resistance (steel) with high thermal conduction (Cu).

Wire Explosion Coatings were first produced by Faraday last century. W, Mo and piano wire onto steel and Al have 5 times the adhesion of flame sprayed versions, due to a welded zone. Mo coating on the inside of Al alloy cylinders of internal combustion engines by this method is 5 times better than the usual Cr plating. The wire vapour reaches the substrate first, which excludes air and prevents oxidation of the immediately subsequent shower of W and Mo molten droplets. The inner surfaces of tubes of 1 cm diameter can be thus coated, to about 10 microns thick from a single explosion which can be built up by repetition. The coating is smoother than by flame or plasma spraying, with high wear resistance and low friction.

Liquid Metal Spraying. The Osprey metal spray process can build up very thick (e.g. 5 cm) metal coatings very fast (minutes). The coating metal is induction melted and atomized by an inert gas annular blast at the bottom of the melt crucible as it emerges through a hole into a reduced pressure area containing the substrate.

Coating by Welding (Hardfacing, Weld-cladding)

Properties of hardfacing alloys are given in Reference 1. Welding methods used are: Gas, Powder, Manual Metal Arc, Metal-Inert Gas (MIG), Tungsten-Inert Gas (TIG) (which includes plasma arc), Submerged Arc and Friction Welding. These methods give a coating which resists heavy wear under mechanical thermal shock. Generally, ferrous alloys are used for abrasion resistance, Ni and Co alloys for oxidation and corrosion resistance, Cu alloys for bearings, and WC, CrB, etc for wear resistance. The main parameter is the dilution which occurs of the coating as some of the substrate melts into it. This can be kept as low as 5%. Cracking of thick layers can be compensated for by applying an intermediate soft metal layer and by pre-heating, especially when brittle and hard alloys are deposited.

Cladding includes explosive impact, magnetic impact bonding, hot isostatic pressing (HIP), and, coextrusion.

Rolling and Extrusion have been used for many ferrous alloys, Ni/Fe alloys, Al alloys, etc for corrosion and oxidation resistance for ambient and high temperature applications in S- and Cl- containing environments and for other high temperature uses such as energy, aviation and nuclear industries. Type 310 stainless steel co-extruded with 50Ni-50Cr alloy has given very good service in hot corrosion conditions in boiler systems. A thin interlayer of electrodeposit usually gives a better bond for a creep resistant steel coated with an overlay cladding.

Explosive Cladding. This can be used (e.g.) to form multilayer laminates of identical alloy sheets like Al-Li which are impossible to diffusion bond.

Diffusion Bonding. Many coating processes include heat treatment as a step to ensure good bonding. But this section describes diffusion bonding as a primary process. The components are pressed together and heated to cause interdiffusion. This is slow. Composition, microstructure, physical, chemical and mechanical properties are affected and controlled by the time of heat treatment, the temperature and pressure and the cooling rate. The temperature should be 0.5 to 0.8 of the m.p. in Kelvin. Additional layers of coatings of a few microns, or loose shims, are interposed either to promote bonding or to act as diffusion barriers to specific elements.

Some disadvantages are:
- poor adherence over large areas due to impurities such as oxides or grit at the interface;
- poor bonding due to interface separation via vacancy concentration and void formation during diffusion;
- unknown thermal effects on the properties of the bonding elements.

Aluminizing and chromizing on steels and Ni alloys have been industrial diffusion bond techniques for decades. The two problems for good bonding of Al on steel are the oxide which forms easily on Al and brittle intermetallics (e.g. $FeAl_3$) which form at the interface. Several remedies are available.

Hot Isostatic Pressing (HIP) is fairly new and is good for complex shapes. Gas turbine alloys have been HIP clad with MCrAlY series coatings and metal/ceramic coating systems have been produced. The coating material is positioned by spot welding or powder spraying and is then enclosed in a flexible membrane (e.g. glass chips which later melt) within a deformable metal container. HIP is then applied at high temperature and pressure and diffusion bonding occurs.

Electromagnetic Impact Bonding uses magnetic energy to give the impact which clads in a similar fashion to the chemical explosion method. In a few microseconds a pressure of 300 MPa can develop. The bonding can be done either hot or cold. But it is limited to circumferentially continuous shapes like tubes and rings. It has been used for gas turbine blades.

REFERENCES

1. M.G. Hocking, V.V Vasantasree and P.S. Sidky, "Metallic and Ceramic Coatings - Production, High Temperature Properties and Applications," Longman (UK) and J Wiley (USA), 1989.
2. L.C. Klein, Ed., "Sol Gel Technology," Noyes Publ. (USA), 1988. See also R. W. Jones, "Fundamental Sol Gel Technology," Ceramic Developments Ltd., Unit A, Marks Rd., St James Indust. Estate, Corby NN18 8AN, UK.

2.1.2
PVD Protective Coatings for Wear Parts

M. Berger and P. Stokley

BALZERS LIMITED, FL-9496 BALZERS, LIECHTENSTEIN

1 INTRODUCTION

More demanding performance criteria, reduction of lubricants and additives and increasingly stringent requirements regarding the reliability of machinery, mean that individual components can be subjected to exceptional loads, wear and corrosion. The problem is all the more critical in the case of a precision component, where a few microns of wear suffice to adversely affect the operability and service life of the machinery.

In order to meet the demands of highly loaded precision parts, functional faces need surface treatments with high wear resistance, high accuracy and high load-carrying capability (Figure 1).

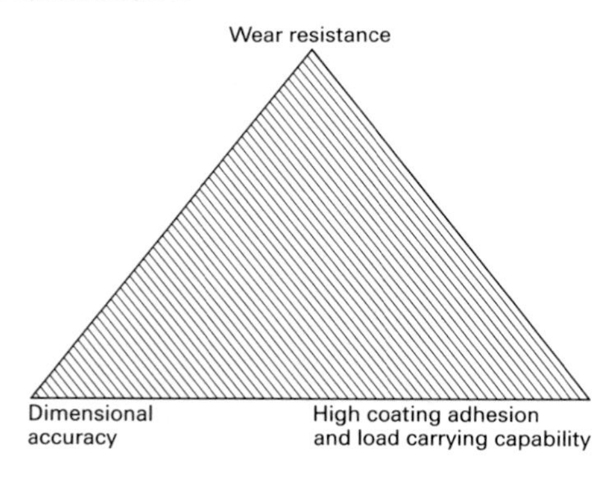

Figure 1 Coating demands for highly loaded precision parts

Conventional processes (Figure 2) apply thick coats (e.g. galvanic coatings or thermal spraying) or need high temperatures (e.g. thermochemical treatment or CVD coating) and consequently they cannot always meet the requirements for dimensional accuracy. In addition neither wear resistance (i.e. hardness and friction coefficient) nor load-bearing capability are always satisfied.

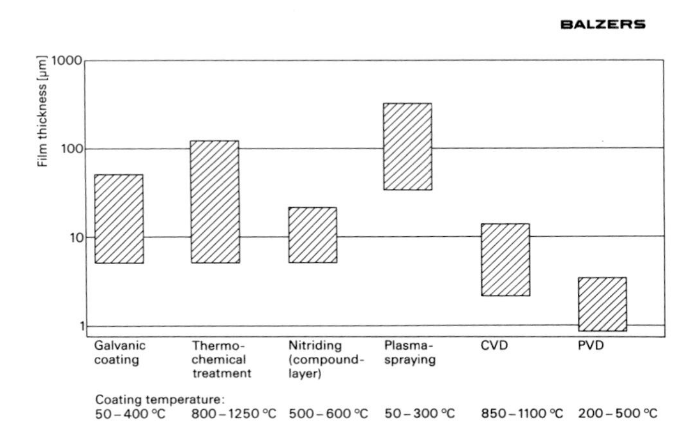

Figure 2 Accuracy of coating

2 PVD COATINGS

PVD coatings are generally between 1 and 4 µm in thickness and are applied between 200 and 500^{0}C. This means that precision components maintain high dimensional accuracy even after coating. Besides high hardness and low friction coefficients PVD coatings show very high load carrying capability, high toughness and excellent adhesion, so that they do not spall even under plastic deformation of the surface. After bending a PVD coated wire, no coating failure is visible (Figure 3). The high toughness and adhesion are due to a metallurgical bonding between substrate and coating (Figure 4). Transmission electron micrographs and elemental analysis show an interfacial layer of some nanometres consisting of substrate and coating materials. This interfacial zone is produced in the first stage of the PVD process (Figure 5). The surface of the workpiece is bombarded with noble gas ions in order to remove any contaminants and to sputter

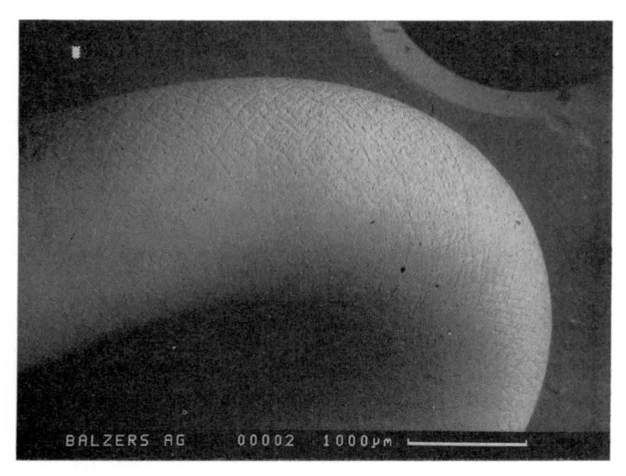

Figure 3 Bend test of PVD coated wire, demonstrating coating/substrate adhesion

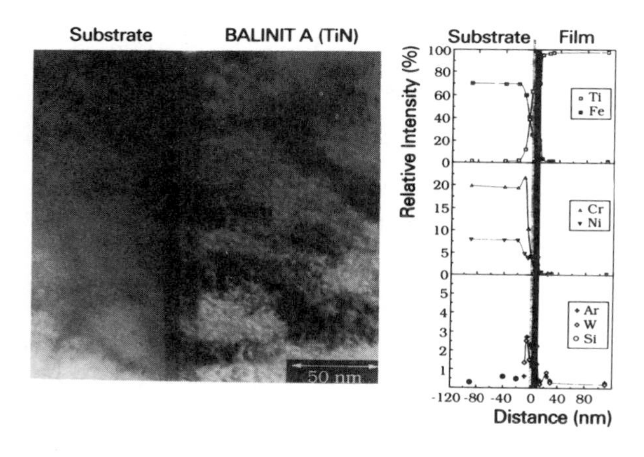

Figure 4 TEM micrograph and elemental profile through the coating/substrate interface of Figure 3

off some substrate material. These substrate atoms condense with the coating material (e.g. titanium) which is evaporated in the second step of the process. Reactive gas (e.g. nitrogen), introduced into the chamber, combines with titanium ions on the surfaces of the workpieces to form hard coatings (e.g. TiN).

Today TiN, TiCN, WC/C and CrN coatings are commonly used to coat precision parts (Figure 6).

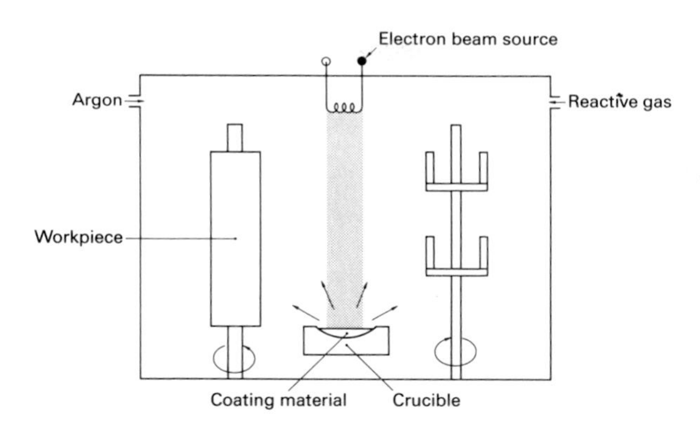

Figure 5 PVD process

BALZERS

	BALINIT A (TiN)	BALINIT B (TiCN)	BALINIT C (WC/C)	BALINIT D (CrN)
Microhardness (HV 0.025)	2300	3000	1000	1750
Coefficient of friction against steel (non greased)	0.4	0.3	0.2	0.5
Film thickness (μm)	1...4	1...4	1...4	1...4
Thermal stability (°C)	600	400	300	700
Resistance against abrasive wear	+ +	+ + +	+	+ +
Protection against cold weldings	+ +	+ +	+ + +	
Resistance against fretting	+		+ +	
Colour of coating	gold-yellow	blue-grey	black-grey	silver-grey

Figure 6 PVD coatings

The industrial use of titanium nitride coatings to protect precision tools against wear dates back to 1980. The material is extremely hard, has good friction-reducing properties and is resistant to high temperatures. It is the ideal coating for precision parts in applications demanding resistance to abrasive-adhesive wear.

A titanium carbonitride (TiCN) coating is the first choice in applications requiring maximum resistance to abrasive wear, because of the high hardness.

The tungsten carbide/carbon (WC/C) coating is applied at temperatures lower than 250^0C and consequently, it can be used to coat parts made of case-hardened or ball-bearing steels. Since the dry friction coefficient is lower than 0.2 (compared to a value of 0.8 for a steel/steel couple) this coating is the first choice in applications requiring maximum resistance to cold welding (e.g. under bad lubrication conditions such as dry running).

Chromium nitride (CrN) coatings have high thermal and corrosion resistance.

In the following applications (i.e. TiN for screw spindles, TiN for gudgeon pins and WC/C for gears) the theoretical coating properties should be compared with results from practice.

If a spindle pump screw is used to pump contaminated oil or grinding emulsion, the pump spindle is exposed to combined abrasive and adhesive wear (Figure 7). Cold welding and grooves are visible on the screw flanks. The pump's service life can be extended significantly by applying a TiN coating to the spindle (Figure 8). The improvement is due to the good resistance of TiN against abrasive- and adhesive wear.

Figure 7 Wear of a spindle pump screw

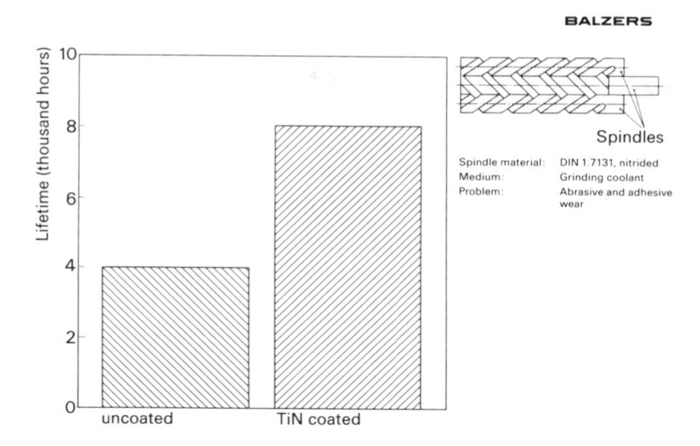

Figure 8 Life time of TiN coated screw of spindle pumps

Abrasive-adhesive wear also occurs on gudgeon pins in combustion engines. Normally sufficient sliding properties are obtained with bronze-brushes. In a new combustion engine, using rape seed oil extremely high combustion pressures are used. The pressure is so high that any kind of bronze bush is plastically deformed by the gudgeon pins. The motor did not reach sufficient service life even at pressures of 100 MPa (Figure 9). In the next development step the piston pins were borided and the bush was removed. The motor failed after 150 MPa. Only with TiN coated pins running directly on the connecting rod the required pressure of 200 MPa is reached for the total service life of 300,000 km.

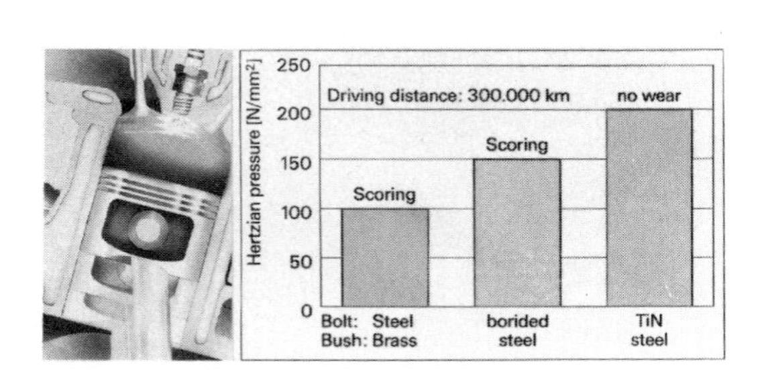

Figure 9 Load-carrying capability of TiN coated gudgeon pins

Completely different wear mechanisms occur at gear flanks. Gears operating under severe loads are subjected to cold welding (i.e. adhesive wear) and pitting formation. Due to the low friction coefficient the resistance against cold welding is increased. But the more severe wear mechanism is the pitting formation (Figure 10). It occurs under high loads or after long working periods. The damage starts on surface notches or little inclusions under the surface by crack formation under the cyclic load of a running gear. Growing cracks cause particle flake off leading to the typical pitting structure. The damage probability of pitting failure is drawn in Wöhler-curves. Figure 11 shows the 50% damage probability of a case hardened gear and WC/C coated gear. The service life is increased by a factor of 2 corresponding to an increased load-bearing capacity of 10 to 15%. The improvement is due to the excellent running-in and smoothing properties of the WC/C coating, lowering the actual hertzian pressure.

Figure 10 Pitting formation on gear flanks

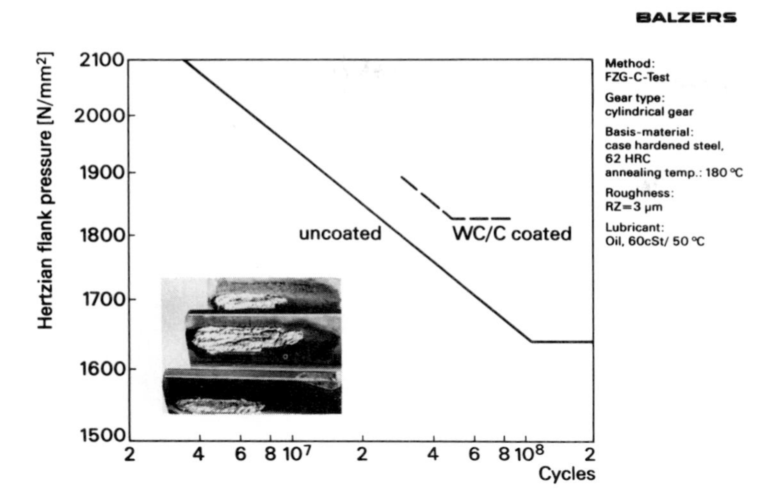

Figure 11 Load-carrying capability of WC/C coated gears

3 CONCLUSIONS

Practical experience has established that if wear of a few microns is sufficient to impair the service life and operability of a machine, PVD coatings effectively protect susceptible precision components (Figure 12). Successful applications were found for hydraulic pumps, pneumatic valves, sliding parts, plastic processing, injection pumps, ball bearings and many other precision parts. The suitability of PVD coatings derives from their high wear resistance, the low defined increase in thickness and the high load-bearing capacity.

Figure 12 Applications of PVD coated precision parts

2.1.3
Characterisation of Erosion – Corrosion Regimes

M.M. Stack, F.H. Scott, and G.C. Wood

CORROSION AND PROTECTION CENTRE, UMIST, PO BOX 88,
MANCHESTER M60 1QD, UK

1 INTRODUCTION

Wastage due to the synergistic effects of erosion and corrosion at elevated temperatures has become of increasing interest in recent years due to the prevalence of such problems in energy conversion and other such industrial processes. For example, erosion-corrosion of materials can occur in environments ranging from coal-conversion processes to catalytic cracking in oil and gas separation to the turbine blades of jet engines. The extent of wastage in these environments is dependent on a wide range of variables which include properties of the impacting particle, the target and the corrosion environment. Several wastage regimes have been identified which correspond to conditions where erosion of the base material, or erosion of the corrosion product layer, is the predominant wastage mechanism. The transition between these regimes has been termed "erosion-corrosion dominated" and marks the temperature above which the wastage rate is significantly greater than for "erosion-dominated" conditions.

Considerable laboratory research in the area has been recently directed at wastage of alloys in laboratory apparatus which simulate fluidized-bed environments in coal- conversion systems. One of the most interesting results from these studies has been the effect of temperature. The wastage rate increased up to a critical temperature and, subsequently, decreased with increasing temperature above this temperature(1-3). The reduction in wastage rate above a critical temperature is attributed to the attainment of a critical oxide thickness(4). This marks the transition to "corrosion-dominated" behaviour where wastage occurs predominately due to brittle chipping of the scale. This peak in the wastage versus temperature curves has been shown to move to higher temperatures with increasing alloy oxidation resistance(1), and to higher temperatures and wastage rates with increasing velocity(5). The effect of increasing oxidation resistance of the substrate is to reduce the amount oxide formed in the time interval between successive erosive events. Hence, the critical oxide thickness is attained at a higher temperature for the alloy with better oxidation resistance. Increasing particle velocity reduces the time interval between impacts and increases the impact energy of the erodent. Hence the critical oxide thickness necessary to effect a transition between "erosion-corrosion" dominated behaviour and "corrosion-dominated" behaviour is attained at successively higher temperatures as the velocity is increased(5), when the particle flux is independent of velocity.

Erosion-corrosion results can be presented on a map to chart the transition between the regimes with changes in the main erosion-corrosion variables such as velocity and temperature. It has been shown how laboratory results on weight change data and surface morphologies can be used to establish the approximate boundaries on such a map(6,7). However, a computer program has been written to establish a more theoretical method of determining the boundaries, using the latest theories proposed to account for the transitions through the regimes as functions of the main variables. The transitions are initially depicted on schematic diagrams representing the effects of temperature and velocity. Subsequently, the initial map established by the program is shown and the effects of increasing oxidation resistance and particle flux are evaluated for the various boundaries. The anticipated changes in corrosion kinetics in the different regimes are also indicated, and future development of such maps is outlined.

2 RESULTS

2.1 Methodology of simulation program

The simulation program divides erosion-corrosion into four regimes which are:

(i) erosion-dominated
(ii) erosion-corrosion-dominated
(iii) corrosion-dominated 1
(iv) corrosion-dominated 2.

Regime (i) describes the situation where erosion of metal is the dominant process. As the oxidation rate increases exponentially with increasing temperature, eventually the weight loss increases more than would be anticipated if erosion of metal were the dominant process, i.e. than could be attributed to the reduction in yield strength of the metal, as the temperature is increased. This marks the transition to regime (ii), which is "erosion-corrosion-dominated". The wastage begins to increase rapidly with further increases in temperature as the scale which forms between two successive erosion events is removed down to the scale-metal interface. However, at a critical scale thickness, the scale is sufficiently thick not to be removed down to the scale-metal interface, between these erosion events. This marks a peak in the wastage rate versus temperature curve and the transition to "corrosion-dominated" behaviour. In previous work, the maps which have been outlined have included two boundaries, marking the transitions between regimes (i) and (ii) and between (ii) and (iii)(6-9). However, for the method to be of practical use to engineers, the conditions for which low wastage at higher temperatures occur must be known. Hence, the corrosion-dominated regime has been divided into two regimes. "Corrosion-dominated 1" refers to the point at which the critical oxide thickness is attained for the conditions, and "corrosion-dominated 2" to the temperature at which the overall weight change is zero, as the overall weight loss decreases with increasing temperature above the critical temperature. These transition points are indicated schematically on Figures 1 and 2 which show the weight loss as functions of temperature and velocity.

<u>2.2 Mathematical algorithms used to calculate the transition boundaries</u>.

The transition temperature (a) between "erosion-dominated" and "erosion-corrosion-dominated" behaviour is attained as the temperature is increased when:

$$W_{er} > W \quad (1)$$

or $$W_{er} > W_{rt}.(1+x)$$

where:

W_{er} = weight loss in a given time interval.

W_{rt} = weight loss during erosion at room temperature in the same time interval as above.

x = % increase in erosion rate of material due to reduction of yield strength(calculated from erosion models) as the temperature is increased.

W = weight loss due to erosion of metal at the erosion temperature in a given time interval.

Since erosion is inversely proportional to yield strength(10), this means that when the measured erosion rate increases are in excess of this value, the damage must involve removal of surface scale, not just loss of metal. This marks the transition to erosion-corrosion dominated behaviour.

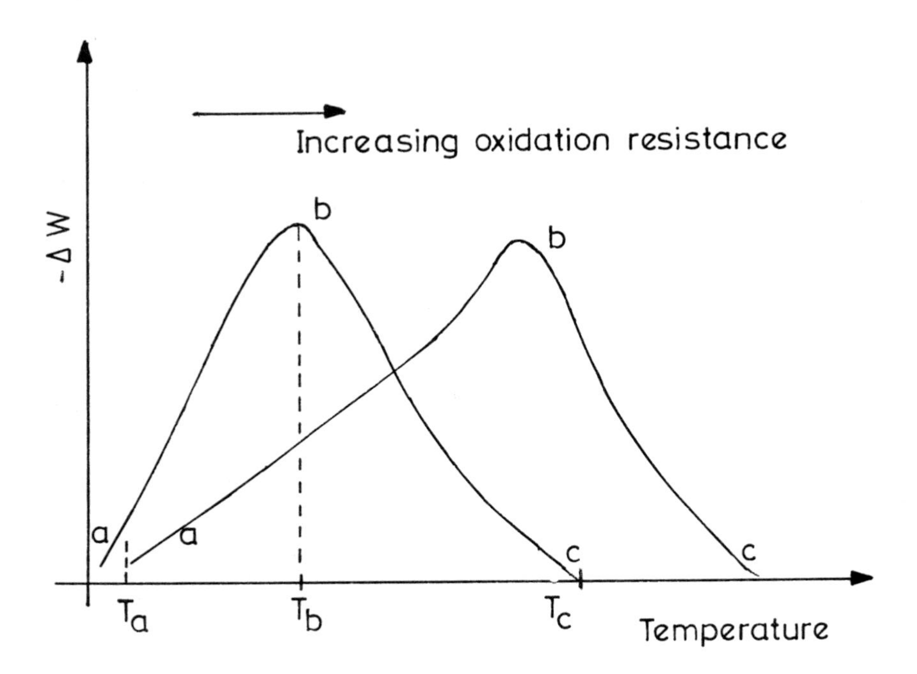

Figure 1 Schematic diagram of weight loss versus temperature for alloys of different oxidation resistance exposed to erosion at high temperature. T(a): Temperature at which transition from "erosion-dominated" to "erosion-corrosion-dominated" behaviour occurs; T(b): Temperature at which transition from "erosion-corrosion-dominated" to "corrosion-dominated 1" behaviour occurs; T(c): Temperature at which transition from "corrosion-dominated 1" to "corrosion-dominated 2" behaviour occurs

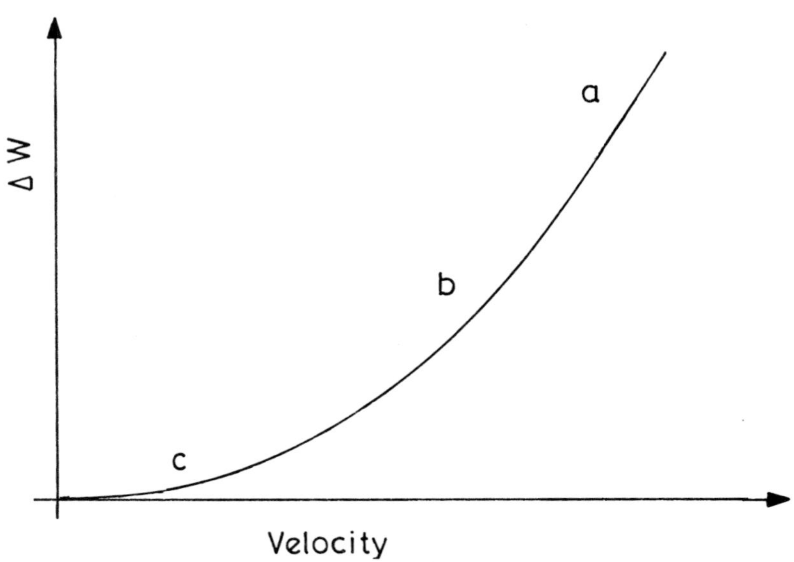

Figure 2 Schematic diagram of effect of velocity on the transition through the erosion-corrosion regimes at a given temperature

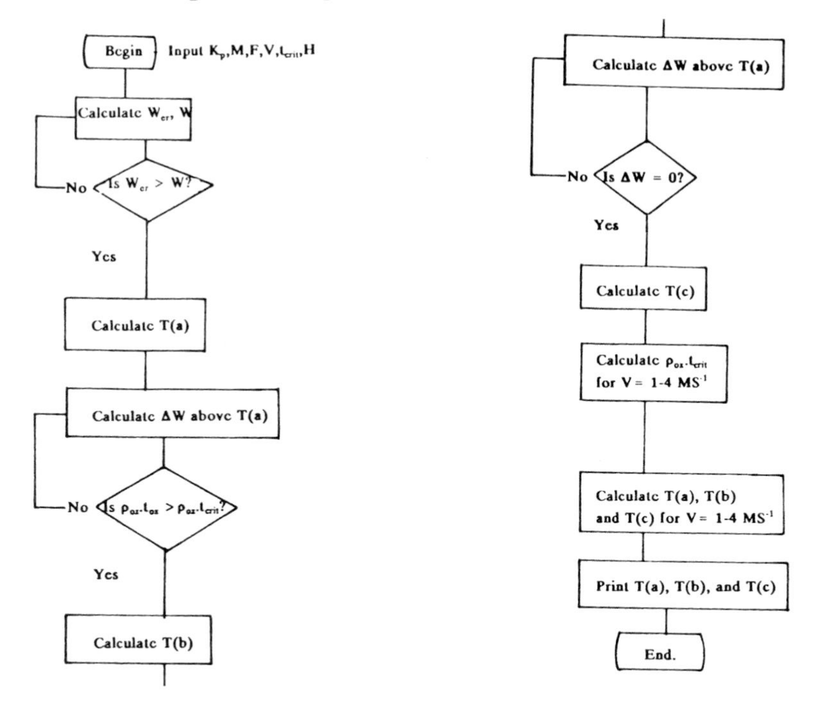

Figure 3 Outline flow chart of simulation program used to calculate boundaries on the maps

As the temperature is increased the oxidation rate increases, i.e.

$$K_p = K_o.\exp(-K/T) \qquad (2)$$

where:

K_p = parabolic rate constant
K_o, K = constants for a given oxide(11)
T = temperature.

The weight gain per unit area as a function of time for oxide growth is given by:

$$W_{ox}^2 = K_p.t \qquad (3)$$

where:

W_{ox} = weight gain
t = exposure time

The weight gain per unit area due to oxidation in a given time interval, t, is also given by:

$$W_{ox} = \rho_{ox}.t_{ox} - f.\rho_{ox}.t_{ox} \qquad (4)$$

where:

ρ_{ox} = density of scale
t_{ox} = thickness of scale
f = atomic weight of metal/molecular weight of oxide($\approx$0.7 for iron based oxides)

If the scale formed between impacts is removed down to the scale-metal interface, the weight loss due to erosion is:

$$W_{er} = \rho_{ox}.t_{ox} + W \qquad (5)$$

Since the contribution of loss of oxide scale to the overall weight loss compared to that of loss of metal increases as the temperature is increased, at a sufficiently high temperature, it can be assumed that, approximately,

$$W_{er} \approx \rho_{ox}.t_{ox} \qquad (6)$$

=> The overall weight change between impacts as the temperature increases is:

$$\Delta W = W_{ox} - W_{er} \qquad (7)$$
$$= \rho_{ox}.t_{ox} - f.\rho_{ox}.t_{ox} - \rho_{ox}.t_{ox}$$

When t_{ox} reaches the value of t_{crit}(the critical oxide thickness), the thickness of the oxide removed becomes relatively independent of temperature. The value of t_{crit} represents the maximum scale thickness which can be removed between successive erosive events. If it is assumed that the worst scenario is that this scale thickness is removed as the temperature is increased above the critical temperature: => above the critical temperature:

$$\Delta W = \rho_{ox}.t_{ox} - f.\rho_{ox}.t_{ox} - \rho_{ox}.t_{crit}.$$

This marks the transition boundary (b) between "erosion-corrosion-dominated" and "corrosion-dominated 1" behaviour. As the temperature is increased further, the overall weight loss decreases. The transition (c) from "corrosion-dominated 1"to "corrosion-dominated 2" behaviour is taken to be the point when the overall weight change is zero.

Increasing the velocity of the erodent particles increases the erosion rate by the following equation:

$$E = K_2 . V^n \qquad (8)$$

where:

E = erosion rate
V = velocity
n = velocity exponent
K_2 = Constant.

If K_2 is assumed independent of temperature, the value of n shows an increase with temperature, up to the critical temperature, due to differences in erosion resistances of metal and oxide scale(12). Hence, values of n = 2 at room temperature and n = 3 at the critical temperature have been used to estimate the erosion rates of metal and the value of the critical scale thickness as the velocity is increased. The transition temperatures, T(a), T(b) and T(c), were evaluated using equations 3 and 4. For the analysis, an initial value of t_{crit} was assumed for the conditions, and the change in this parameter with velocity was estimated from equation 7. A flow chart of the main steps in the simulation program is given in Figure 3.

2.3 Simulation results

Figure 4 shows the general features of the erosion-corrosion map evaluated by the above technique. It can be seen that increasing the velocity shifts the boundaries for "erosion-corrosion dominated" and "corrosion-dominated" behaviour to higher temperatures. Similarly, increasing temperature results in a transition through the erosion-corrosion regimes. If the oxidation rate of the target material increases by a factor of 10^2, Fig.5, the boundaries for "erosion-corrosion dominated" and "corrosion-dominated" behaviour occur at lower temperatures than for the original material shown in Fig.4. This simulates the relative erosion-corrosion behaviour which has been observed for alloys of different oxidation resistances i.e. mild steel compared to 347 stainless steel(Fe-18%Cr-10%Ni)(1). The oxidation resistance of the latter is significantly greater and, hence, the transition temperatures for "erosion-corrosion-dominated" and "corrosion-dominated" behaviour are higher for the alloy with the superior corrosion resistance. If the particle flux is increased by a factor of 4, the time interval between impacts is reduced by a quarter. From equation 4, the oxide thickness formed between erosive impacts is halved, and, hence, the temperature at which the critical

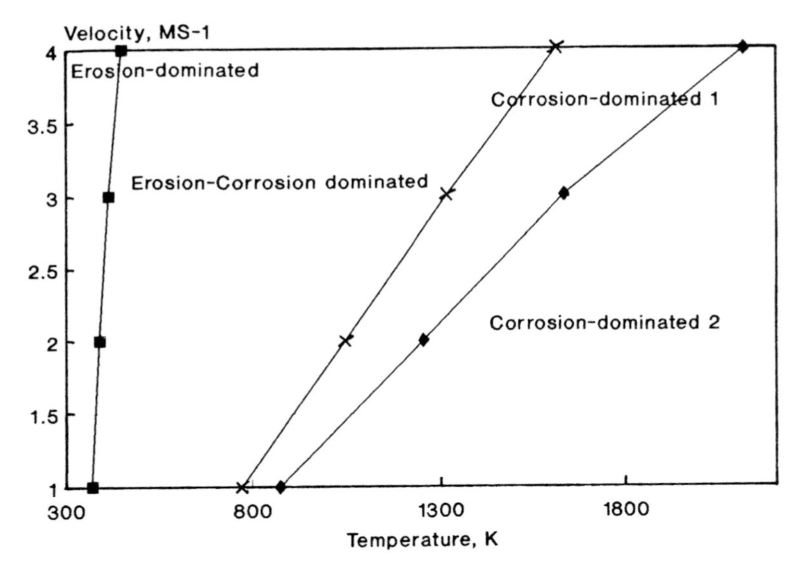

Figure 4 Erosion-corrosion map evaluated by the simulation program

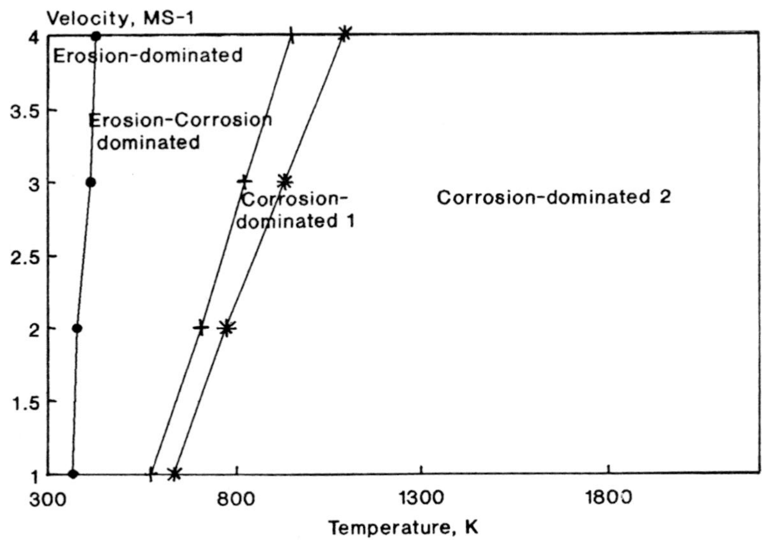

Figure 5 Variation of boundaries on the map when the oxidation rate is increased by 10^2 compared to Figure 4

scale thickness occurs is higher than for the original flux. Thus, the transition boundaries move to higher temperatures as the particle flux is increased.

3. DISCUSSION

Characterization of erosion-corrosion regimes using erosion-corrosion maps is important because the parameters involved in such processes are numerous. Since materials selection in these environment may involve a trade-off of one parameter against another, such as particle concentration and velocity, maps can be used to aid this exercise, and thus can be a useful tool for the process engineer. The above method provides a basis for identifying the important boundaries between the regimes, particularly at high temperatures where transition(c) marks the temperature at which the weight change is very low in comparison to that at intermediate temperatures.

A problem with any such technique is the number of assumptions in the simulation process. This model assumes that oxide adhesion does not change with temperature and composition. However, the literature suggests that this may not always be true. The model also assumes parabolic growth kinetics for simplicity. However, the growth kinetics change with temperature, particularly within the erosion-corrosion-dominated regime as shown in Fig.5. When the scale is removed to the scale-metal interface between erosion events, transient oxidation kinetics will predominate. These may be linear or involve a transition from linear to parabolic, depending on the time between events and temperature. At present, no method has been established to estimate the critical scale thickness for a given material in a given set of conditions, although estimates of this value can be made from the literature(1-4). This is due to the fact that the critical scale thickness may be dependent on a range of parameters which affect scale adhesion. These include ductility of the scale formed and its hardness relative to that of the substrate. Other assumptions have included those on the variation of velocity exponent as a function of temperature. This variation is due to the differences in erosion resistances of scale formed between impacts and the erosion resistance of the substrate, i.e., the K_2 values in equation 7. However, since these parameters are a function of oxide composition in addition to temperature, it is simpler to incorporate this difference in values of n which have been obtained from the literature(12).

Future development of such maps will be to simulate the effect of properties of the target(hardness), the corrosive environment(gas composition, thermal gradient between particle and surface), and particles(mass and corrosivity). Estimation of lifetimes of coatings in such

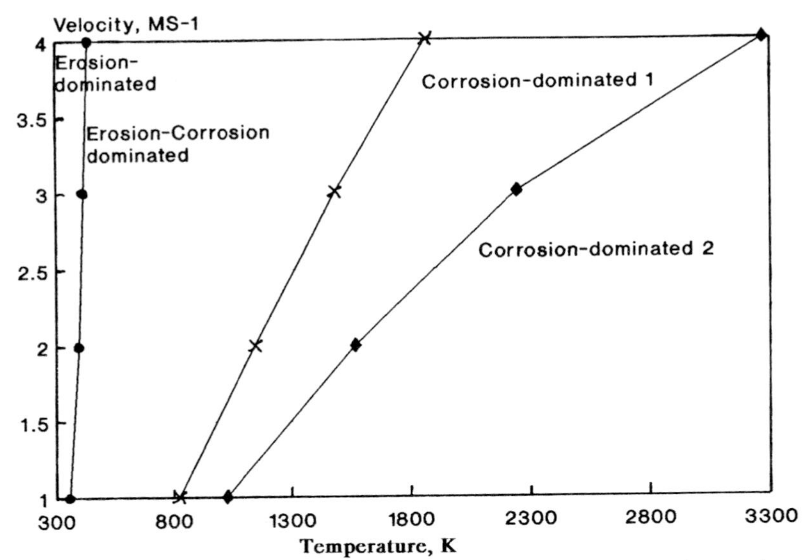

Figure 6 Variation of boundaries on the map when the particle flux is increased by a factor of 4 compared to Figure 4

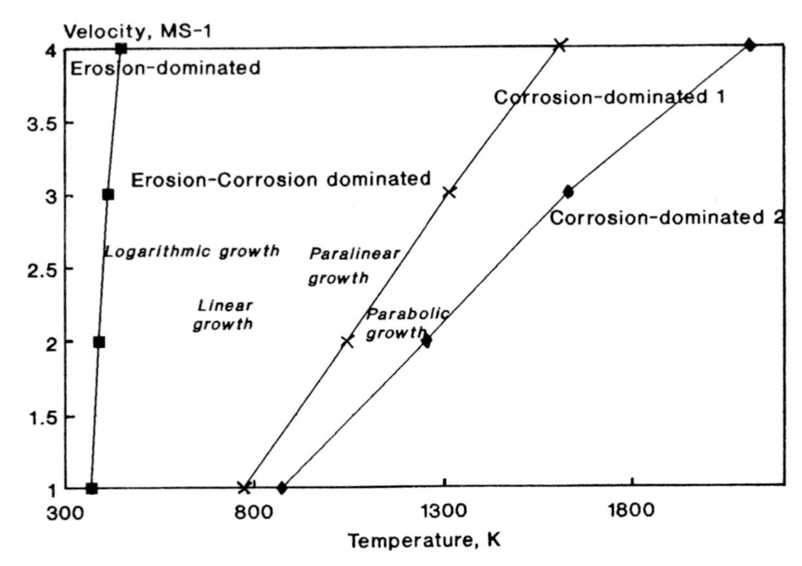

Figure 7 The anticipated change in corrosion kinetics as the transition through the erosion-corrosion regimes take place on the map

environments is important and this phenomenon can be represented on such a map. At present, the interaction of no more than two variables can be represented. The most critical variables in process plant in which erosion-corrosion may occur are particle concentration, particle size and velocity, as these change fundamentally as the gas passes through filters in the plant at elevated temperatures. The development of the map into a three-dimensional form is an important aim of the work, in addition to addressing the factors outlined above.

4. CONCLUSIONS

(i) A simulation program has been written to establish the boundaries between erosion-corrosion regimes as functions of temperature and impact velocity.
(ii) The maps have been used to show the transition between regimes as erodent flux and oxidation rates are increased.
(iii) Future developments will be to incorporate additional variables on the map and representation of the boundaries in a three dimensional format.

5. REFERENCES

1. A.J. Ninham, I.M. Hutchings and J.A. Little, Proc. Conf. CORROSION '89, NACE, Houston, Tx, 1989, paper 544.
2. F.H. Stott, M.M. Stack and G.C. Wood, Corrosion-Erosion-Wear of Materials at Elevated Temperatures, NACE, Houston, Tx, 1990, 12, 1-16.
3. D.J. Hall and S.R. J. Saunders, Proc. High Temperature Materials for Power Engineering, Ed. E. Batchelet, Liege, 190, 1, 157.
4. I.G. Wright and V.K. Sethi, Proc. Conf. CORROSION '90, NACE, Houston, Tx, 1990.
5. V.K. Sethi and R.G. Corey, Proc. 7th Int. Conf. Erosion by Liquid and Solid Impact, Pub. Cavendish Lab. Univ. of Cambridge, Cambridge, 1987, paper 73.
6. M.M. Stack, F.H. Stott and G.C. Wood, Materials Science and Technology, 1991, 7, 1128-1137.
7. M. Stack, F.H. Stott and G.C. Wood, Heat Resistant Materials, ASM, Ohio, U.S.A, 1991, 333-342.
8. M.M. Stack, F.H. Stott and G.C. Wood, Journal of Physics D: Applied Physics, 1992, 5, A170-A176.
9. M.M. Stack, F.H. Stott and G.C. Wood, Proc. Conf. Irish Materials Forum 7, University of Limerick, 1991(in press).
10. I.M. Hutchings, Proc. 5th Int. Conf.on "Erosion by Liquid and Solid Impact",Pub. Cavendish Lab. Univ. of Cambridge, Cambridge, 1979, 36-1.
11. N. Birks and G.H. Meier, "Introduction to High Temperature Oxidation of Metals", Edward Arnold, London, 1983.
12. M.M. Stack, F.H. Stott and G.C. Wood, Journal de Physique, 1992(in press).

2.1.4
Wear Maps for Nitrided Steels

H. Kato and T.S. Eyre

DEPARTMENT OF MATERIALS TECHNOLOGY, BRUNEL UNIVERSITY, UXBRIDGE UB8 3PH, UK

1 INTRODUCTION

Nitriding is a ferritic thermochemical treatment in which nitrogen is diffused into a steel surface, at a temperature usually within the range 480-570°C, and has been very widely used for engineering components to improve wear properties[1-3]. After nitriding, a thin compound layer, called white layer, and a relative thick diffusion layer are produced on the component surface. The surface layers have high hardness, depending on the process temperature and the substrate material. Those elements that form nitrides in steel, such as Al, Cr and Mo, have a significant effect on increasing the surface hardness.

There are many studies on the wear behaviour of gas nitriding[4-6], plasma nitriding[7-10] and salt-bath nitrocarburizing[11-13], which show that nitrided steels have high wear resistance. However, most work has been carried out under narrow ranges of load and sliding speed. Wear mechanisms are, however, affected considerably by tribological parameters such as applied load and sliding speed. Gleave[14] investigated the dry wear behaviour of some ferritic thermochemical treatments on BS970,080M80 steel over a wide range of sliding speed. He showed that the thermochemical treatments almost eliminated severe wear over the whole sliding speed range, contrasting with a mild-severe-mild wear transition behaviour of untreated steel, which was broadly similar to that in work published by Welsh[15]. On the other hand, Whittle and Scott[4,16] have observed a transition from mild to severe wear in untreated and nitrided austenitic alloys. These investigations have provided new information on the wear mechanism of nitrided steel, but have not been extended to wear mapping.

After Welsh's work[15] on the systematic wear transition diagram, wear mechanism maps for steel under dry conditions were proposed by Lim and Ashby[17]. The wear maps are very useful for understanding wear mechanism. Therefore, wear maps of surface treated steel would be also useful for engineers and designers to make the best use of surface treatments and coatings.

The aim of this work is to establish wear maps for nitrided steels based on variations in load and sliding speed, by examining the wear mechanisms. In this paper, the first part of this programme is described, being based on a variation in load.

2 EXPERIMENTAL DETAILS

2.1 Wear tests

A pin on disc machine was used which is capable of testing materials under a wide range of loads and sliding speeds. The wear disc, 12mm thick and 130mm in diameter, was fixed on the rotating shaft, and the wear pin, 35mm long and 12mm in diameter (wear face: flat, 2mm in diameter), was loaded by dead weights via a load arm. The friction force and the linear descent of the pin during a wear test were measured by transducers connected to an ultra violet recorder. All wear tests were carried out under dry conditions at room temperature. As the first step of this programme, the sliding speed was kept constant at 0.5m/s and only the applied load was varied. The effect on the sliding speed will be discussed in a future paper.

2.2 Specimen materials

The materials used for the wear pins were untreated and nitrided BS970,905M39 steels. The chemical composition of the steel is given in Table 1. Nitriding was carried out in a NH_3-gas atmosphere furnace at 520°C for 80 hours. The wear discs were produced from BS970,535A99 bearing steel (Table 1) heat treated to

Table 1 Chemical composition of steels (wt%)

steel	C	Si	Mn	Cr	Mo	Al
BS970, 905M39	0.35-0.43	0.10-0.45	0.40-0.65	1.40-1.80	0.15-0.25	0.90-1.30
BS970, 535A99	0.95-1.10	0.10-0.35	0.40-0.70	1.20-1.60	—	—

a hardness of Hv700. All wear specimens were carefully degreased prior to the wear tests.

3 RESULTS AND DISCUSSION

3.1 Metallurgical observation of the nitrided steel

Fig.1 shows the microstructure of the cross-section of the gas nitrided steel pin. A compound layer (white layer) of 25-30μm thick and a relative thick diffusion layer were produced as usually observed in gas nitrided steel[18]. There is some porosity in the outer part of the white layer, which may be formed by the precipitation of molecular nitrogen due to the high nitrogen pressure within the white layer[19,20]. It was determined by X-ray diffraction using Cu-Kα radiation that the white layer consists of ϵ-$Fe_{2-3}N$.

Fig.2 shows the microhardness profile of the gas nitrided steel. The surface layer has a peak hardness of 1000-1100Hv, and the effective case depth defined by the distance between the surface and the layer having a hardness of 550Hv (B.S.6479:1984) is 0.45mm. The reason for the low hardness observed near the surface is that in the hardness test the indentor caused cracks in the white layer probably due to its brittleness[9,21,22]. In addition, the applied test load of 300g might be too high to be supported close to the edge of the sample.

Figure 1
Microstructure of surface layer of gas nitrided steel pin

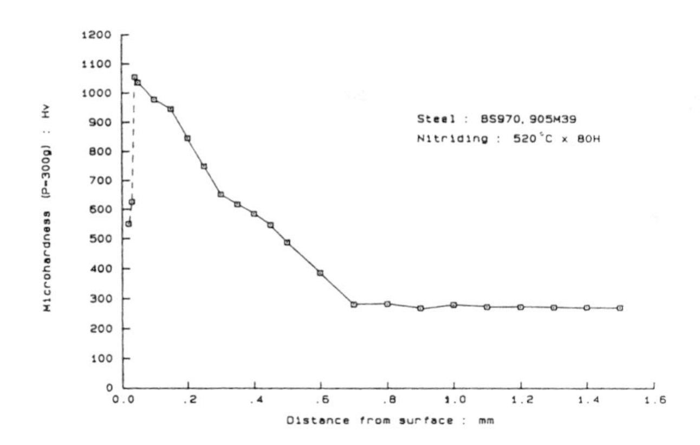

Figure 2 Microhardness profile of gas nitrided steel pin

3.2 Wear test results

The amount of wear was determined by the weight loss of pins. The relationship between the weight loss of the pins and sliding distance at various applied loads is shown in Fig.3. The equivalent amount of weight loss to the white layer depth was so small (only 0.7mg) that the wear behaviour of the nitrided pins was controlled mainly by the diffusion layer. It can be seen that the weight loss of the pin is proportional to the sliding distance except for the untreated pin tests at the load of 5N and 10N in which running-in wear was observed.

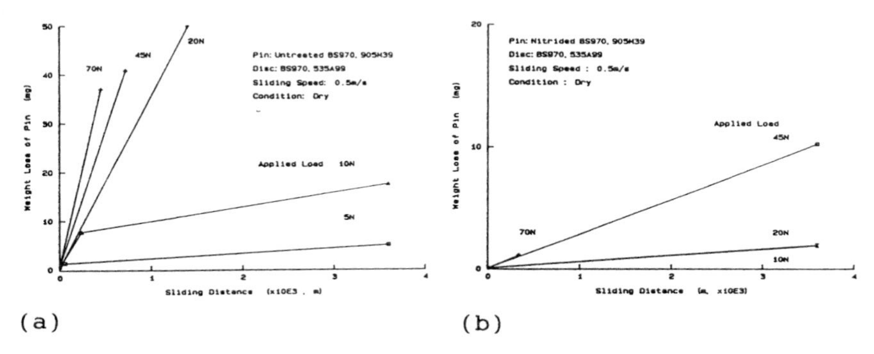

(a) (b)

Figure 3 Relationship between weight loss of pin and sliding distance for (a) untreated pin and (b) nitrided pin

The wear rate obtained from the slope of the straight line at steady state in Fig.3 is plotted against the applied load and shown in Fig.4. It was found that nitriding reduces the wear rate by a factor of four in the low wear regime. Another point to be emphasized in Fig.4 is that a wear transition was observed for both untreated and nitrided steels. This suggests a change in wear mechanism at the transition load. Nitriding increased the transition load from 20N to 45N. In the high wear regime, the wear rate is also reduced by an order of magnitude by nitriding. Welsh[15] has observed a T_1 transition from mild(oxidative) to severe(metallic) wear and a T_2 transition from severe back to mild wear. Therefore, the transitions identified in Fig.4 should correspond to the Welsh T_1 from mild to severe wear. In this work, tests at a higher load than 80N were not carried out because of the high noise level.

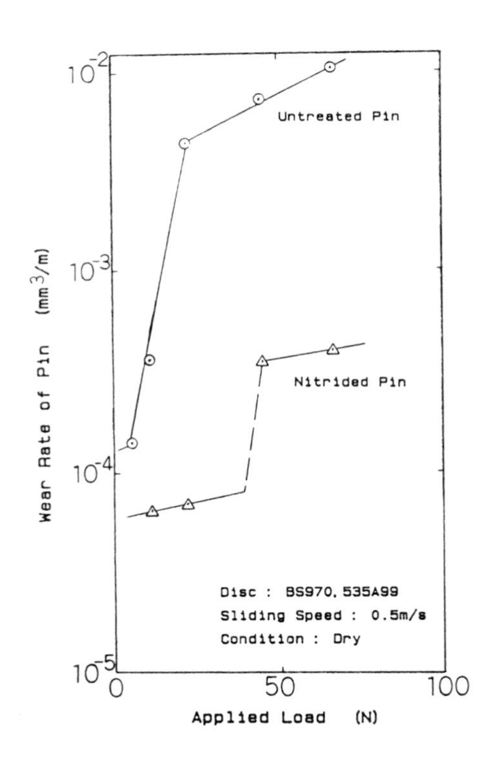

Figure 4 Wear rate of pin at steady state plotted against applied load

3.3 Examination of worn surfaces and wear debris

In order to determine the wear mechanism, the worn surface of the pins and the wear debris were examined with Scanning Electron Microscopy (SEM). Fig.5 shows the surface of the untreated and nitrided pins after the wear tests at the load of 10N (mild wear regime) and 45N (severe wear regime). It can be seen that the untreated pin surfaces are very rough and that heavy plastic deformation occurred at high load(45N). On the other hand, nitrided pin surfaces are much smoother, but there is no significant difference between mild and severe regimes.

Fig.6 shows the SEM of the wear debris generated during the wear tests. The debris from the nitrided pin tests is smaller than that from the untreated pin tests, in addition the debris in the severe regime is larger than that in the mild regime. In the untreated pin test at a load of 45N, large wedge-like debris was

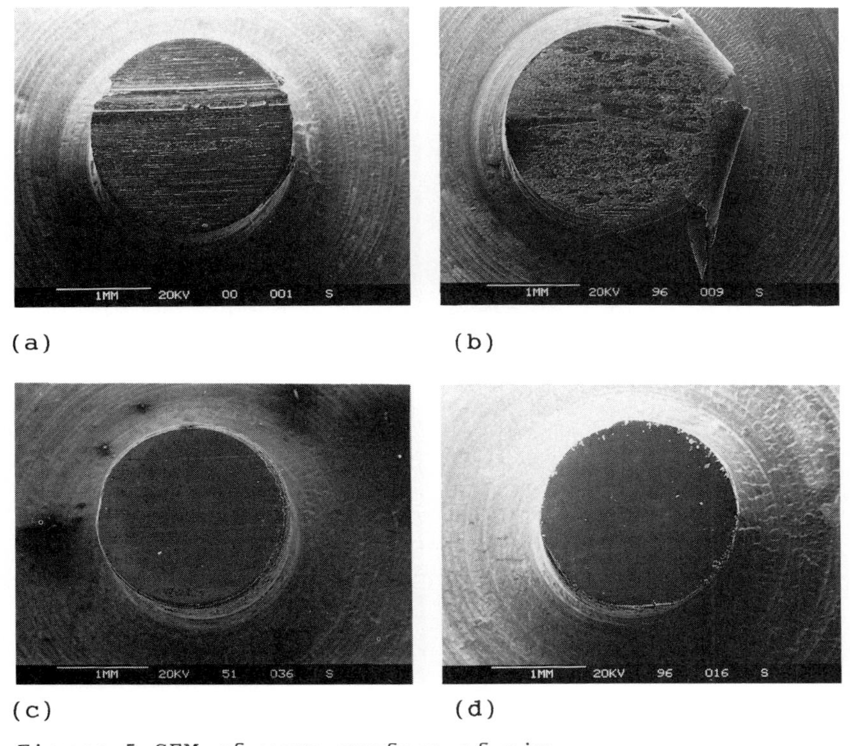

Figure 5 SEM of worn surface of pin
(a)untreated, load:10N (b)untreated, load:45N
(c)nitrided, load:10N (d)nitrided, load:45N

produced, while very fine plate-like and some spherical debris was observed in the nitrided pin test. The results of size and shape of the wear debris are summarized in Table 2.

It is widely recognized that oxide films are formed on surfaces in mild wear, and that these will prevent intermetallic contact and adhesion, resulting in a low wear rate[15,23]. Consequently, the mild wear produces oxide debris, contrasting with the coarse metallic debris generated in the range of severe wear. However, X-ray diffraction studies of the debris indicate that in both mild and severe regimes, Fe and Fe_2O_3 were observed in the untreated and nitrided pin tests, respectively (Table 2). Therefore, this result is contradictory to Welsh's work, while it can be concluded that the low wear rate of the nitrided pins is due to the oxidative wear mechanism. In this investigation, another type of wear mechanism such as delamination proposed by Suh[24], might occur. Further examination is necessary at this stage.

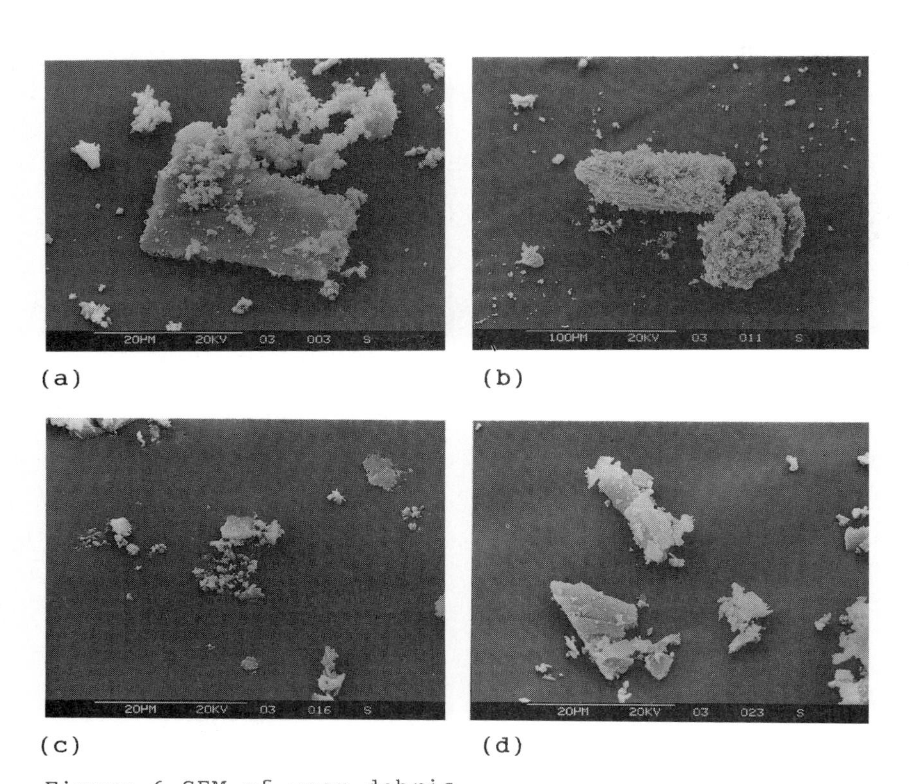

(a) (b)

(c) (d)

Figure 6 SEM of wear debris
(a)untreated, load:10N (b)untreated, load:45N
(c)nitrided, load:10N (d)nitrided, load:45N

Table 2 Results of wear debris analysis

	load:10N (mild regime)	load:45N (severe regime)
untreated pin	metallic (Fe) ~80μm plate-like, spherical	metallic (Fe) ~200μm wedge-like, spherical
nitrided pin	oxidative (Fe_2O_3) ~5μm plate-like, spherical	oxidative (Fe_2O_3) ~20μm plate-like, wedge-like

4 CONCLUSION

Wear mechanism maps for nitrided steels under dry sliding conditions have been proposed. The wear tests at the sliding speed of 0.5m/s have shown that nitriding treatment improved wear resistance by a factor of four at the low wear regime and by an order of magnitude at the high wear regime. A wear rate transition was identified for untreated steel at 20 N and nitrided steel at 45N. Although metallic wear debris was observed in untreated pin tests, oxidized debris was observed in all nitrided pin tests. The effect of sliding speed on wear mechanisms will be discussed in further work to complete the wear maps. These will be useful for designers and engineers to make the best use of the nitriding treatment for further applications. They will also enable comparison to be made with other processes and materials.

REFERENCES

1. J.C. Gregory, Tribology Int., 1978, 11, 105
2. T. Bell, Metallurgia, 1982, 49, 103
3. F. Hombeck and T. Bell, Surface Eng., 1991, 7, 45
4. R.D.T. Whittle and V.D. Scott, Metals Tech., 1984 11, 231
5. T. Burakowski, J. Senatorski and J. Tacikowski, Surface Eng., 1987, 3, 239
6. A. Wells and M.P. Shaw, Wear, 1985, 103, 29
7. K.S. Cho and C.O. Lee, Wear, 1980, 64, 303
8. A. Cohen and A. Rosen, Wear, 1986, 108, 157
9. M.B. Karamis, Wear, 1991, 150, 331
10. T. Bell and Y. Sun, Surface Eng., 1990, 6, 133
11. F.D. Waterfall, Engineering, 1959, 187, 116
12. P.K. Williamson, Tribology Int., 1980, 13, 51
13. R. Marappan and A.R. Rao, Wear, 1988, 121, 239
14. C. Gleave and M. Farrow, 'Heat Treatment '81', 123, The Metals Society, London, 1981
15. N.C. Welsh, Phil.Trans.R.Soc.Lond., A, 1965, 257, 31
16. R.D.T. Whittle and V.D. Scott, Metals Tech., 1984, 11, 392
17. S.C. Lim and M.F. Ashby, Acta Metall., 1987, 35, 1

18. K.E. Thelning, 'Steel and its heat treatment', Butterworths, London, 1984
19. K. Schwerdtfeger, P. Grieveson and E.T. Turkdogan, Trans.Met.Soc.AIME, 1969, 245, 2461
20. P.M. Hekker, H.C.F. Rozendeal and E.J. Mittemeijer, J.Mater.Sci., 1985, 20, 718
21. T. Bell, B.J. Birch, V. Korotchenko and S.P. Evans, 'Heat Treatment '73', 51, The Metals Society, London, 1973
22. D.B. Clayton and K. Sachs, 'Heat Treatment '76', 1, The Metals Society, London, 1976
23. J.F. Archard and W. Hirst, Proc.R.Soc.Lond., A, 1956, 236, 250
24. N.P. Suh, Wear, 1973, 25, 111

2.1.5
Increase of Wear Resistance to Audio Heads by Ion Implantation

K. Wang, Z. An, G. Lu, and G. Wang

DEPARTMENT OF MECHANICAL ENGINEERING, TSINGHUA UNIVERSITY, BEIJING, 100084, PEOPLE'S REPUBLIC OF CHINA

1 INTRODUCTION

Ion implantation is one of the newer surface treatment methods. It has many special advantages. Ion implantation means that the ions of a selected element are implanted on the surface of a component. This method does not affect the degree of dimensional accuracy of the surface finish of the component. There is no distinguishing boundary between the implanted layer and the substrate. There is no problem of separation. The implanted layer has good wear resistance. Therefore this method is used widely in many aspects of science and technology[1-4]. In this paper the method is used for the surface strengthening treatment of audio heads. N^+ or Ti^+ is implanted on the surface of audio heads. Good results are obtained: the implanted layer has good wear resistance, meanwhile the audio heads retain their ordinary electro-magnetic parameters. Ion implantation is beneficial in prolonging the life-span of audio heads.

2 EXPERIMENTAL METHODS

The audio heads used are R4061 double sound gate heads. The head core material is hard permalloy (Ni76Nb6Mo4.5), while the head case is permalloy (Ni74Mo3).

A 400 kW ion implanting machine is used to do the ion implantation. The ions implanted are N^+ or Ti^+. The implanting mode is scanning implantation. The implanting energy is 80 keV. The quantity of implantation is $2x10^{17}/cm^2$. The density of the ion beam current is 10 mA/cm^2. The system vacuum is $9x10^{-4}$ Pa during ion implantation.

The electro-magnetic parameters and microhardness before and after ion implantation are tested. The wear testing of the audio head is carried out to both ion implanted and untreated audio heads in order to compare their wear resistance.

The experimental set-up of the wear test is composed of automatic counters and LX401 cassette recorders which have auto reverse. The experimental conditions are the same as the user conditions for the audio heads. The experimental tapes are γ-Fe_2O_3 audio tapes.

The experimental procedure corresponds to the Chinese State Standard[5]. The following measures and analyses were carried out at every stage of the wear test:

1. measure electro-magnetic parameters of audio heads, including impedance, sound emission sensitivity, sound emission frequency characteristics, recording and playing sensitivity and frequency characteristics;
2. microhardness of the audio heads was measured by HX-1 microhardness meter;
3. a Taylor-Hobson surface appearance instrument was used to measure wear appearance at the audio head slot;
4. S-450 SEM was used to observe the wear appearance of the audio head.

3 EXPERIMENTAL RESULTS AND ANALYSES

The Electro-Magnetic Parameters of the Audio Heads Before and After Ion Implantation

After the audio head undergoes ion implantation, its electro-magnetic parameters may change. Nevertheless the variation of the parameters (frequency characteristics) cannot be more than 6 dB, otherwise the audio head is said to have failed and it cannot be used.

The results of the electro-magnetic parameters measured before and after implanting N^+ or Ti^+ are shown in Tables 1 and 2. It can be seen that the electro-magnetic parameters of the audio heads do not change much after implanting N^+ or Ti^+. Therefore it is possible to increase the wear resistance of the audio heads by ion implantation.

The electro-magnetic parameters of audio heads depend mainly on the head slot. The dimension of the head slot does not change due to the low temperature change during implantation.

Table 1 Electro-magnetic parameters of audio heads before and after N^+ implantation

Electro-magnetic Parameters		Before Implanting			After Implanting		
		1	2	3	1	2	3
Emission freq.,dB	1 Gate	9.6	8.5	10.1	10.0	9.0	9.5
	2 Gate	9.1	10.0	10.2	9.5	10.0	10.0
Rec. pla.freq dB	1 Gate	3.5	4.5	4.0	3.5	4.0	4.5
	2 Gate	3.5	4.0	3.5	3.0	4.0	3.5

Table 2 Electro-magnetic parameters of audio heads before and after Ti^+ implantation

Electro-magnetic Parameters		Before Implanting			After Implanting		
		4	5	6	4	5	6
Emission freq.,dB	1 Gate	9.6	9.0	10.0	10.0	9.5	9.5
	2 Gate	9.5	10.0	9.0	10.0	9.5	9.5
Rec. pla.freq dB	1 Gate	3.5	4.0	4.5	3.0	3.0	2.5
	2 Gate	3.5	3.5	4.0	2.0	2.5	3.5

The treatment method does not form a non-magnetic layer on the surface of the head and it does not affect the magnetic characteristics of the audio head[6]. Therefore the electro-magnetic parameters do not change much after ion implantation.

The Microhardness of Audio Heads Before and After Ion Implantation

The microhardness of audio head component parts before and after ion implantation was measured using a microhardness meter. The results measured are shown in Table 3. It is obvious that the microhardness of audio head component parts increases after ion implantation.

Originally the microhardness of the shielding case and plates is lower than that of the head core. Their hardness increases more than the core's after ion implantation. The difference in microhardness between the core and the shielding case decreases.

It should be pointed out that the hardness shown in Table 3 does not reflect completely the hardening effect of ion implantation and the practical hardness of the implanting layer is higher than that measured. The reason is that the implanted layer is very thin and the maximum implanting depth is about 2000Å. The depth of the microhardness indent is of the order of microns which is much deeper than that of the implanting layer.

Table 3 Hardness of component parts of audio heads ($Hv_{0.01}$)

Component parts	Before implantation	After implantation
Head core	216.5	272.7
Shielding plate	145.2	195.6
Shielding case	158.6	235.4

There are several reasons to explain the increase of microhardness[3,7]: the saturant solid solution is formed after the ions enter the surface of the audio head; vacancies, gapping atoms and a high density dislocation net are formed due to damage by radioactive bombardment of the implanting layer. The increase of hardness is beneficial in enhancing wear resistance of the audio head.

Wear Test

Wear Resistance of Audio Head. Figure 1 is a wear appearance curve of an audio head after working 1000 hours. The X axis is the width of the head core and shielding case. The Y axis is the wear depth. The left half and right half of the curve are the wear appearance curve of the head core and the shielding case respectively. In Figure 1, ab and cd correspond to the width of the two sound gates. The distance between ab and

AB, cd and CD is the average wear depth of the two sound gates. The wear depth of the shielding case can be obtained similarly. The same approach may be used to compare wear resistance of the audio head before and after ion implantation.

The wear depth of the head core and the shielding case versus the tape-travelling time is shown in Figure 2 and Figure 3. It can be seen that the wear depth of the core and the shielding case increases linearly with the increase of the tape-travelling time. After N^+ or Ti^+ ions are implanted, the wear depth of the core and the case decreases and so the wear resistance of the audio head increases. Note the wear resistance of the head increases more after Ti^+ ions are implanted. Likewise, the difference in the wear depth between the core and the case is larger before ion implantation and the inhomogeneity of the wear is severe. After the ion implantation, the wear resistance of the case increases more than that of the core. The wear depth of the case decreases greatly. The difference in the wear depth between the core and the case decreases and so the wear inhomogeneity improves.

The specific wearability (wear depth per hour) of the core and the case, which is calculated in order to clearly explain the increase in wear resistance, is shown in Figure 4. After ion implantation, the specific wearability decreases and the increase of wear resistance is over 100%. The difference in the specific wearability between the core and the case is $6x10^{-6}$ mm/h before ion implantation. That difference reduces to $2.5x10^{-6}$ mm/h after ion implantation. The wear inhomogeneity is improved greatly.

The increase in wear resistance and the improvement of the wear inhomogeneity are related to the increase in hardness after ion implantation.

The thickness of the ion implanting layer is only 2000Å, but the effects of ion implantation reach over 10µm. This indicates that implanted elements can transfer to deep layers during the wear process. This phenomenon is also found in other materials. The current explanation of the phenomenon is as follows[3,7,8]: during the wear process, the contact wear stress has cold-work hardened the materials, dislocation nets are generated continually and the nets extend to the deep locations in the materials.

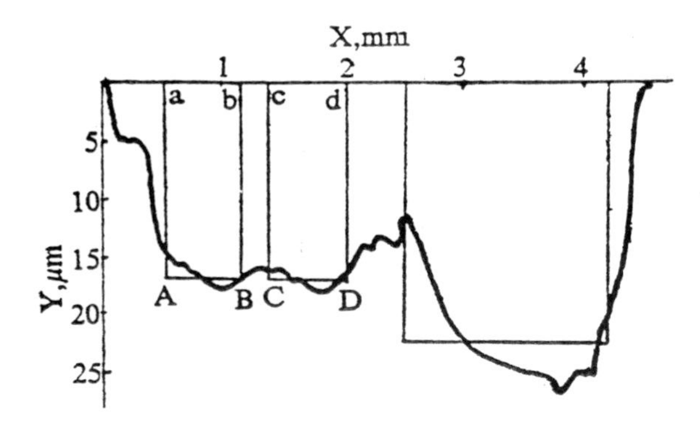

Figure 1 Wear appearance curve of an audio head

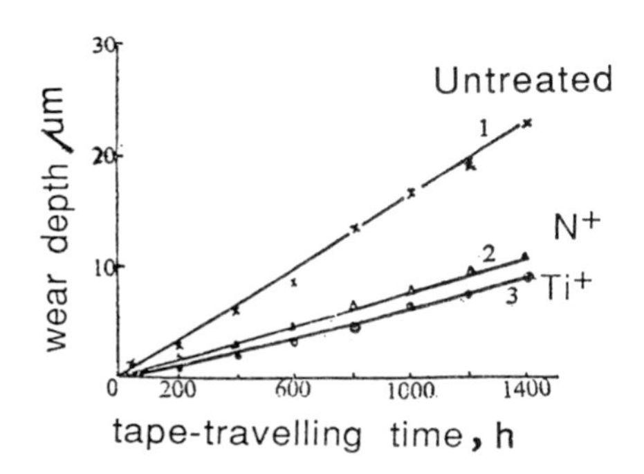

Figure 2 Wear depth of core vs. tape-travelling time

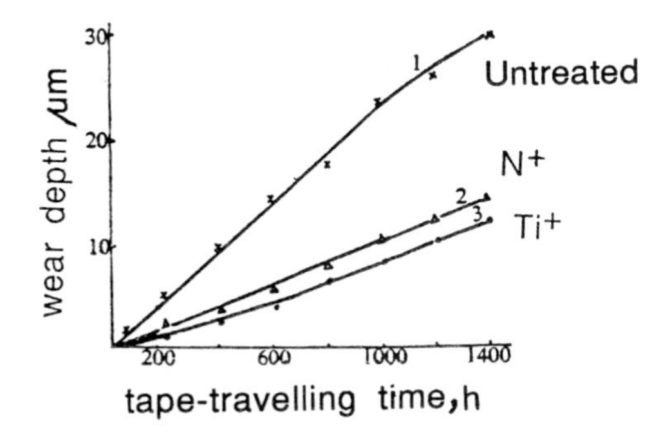

Figure 3 Wear depth of case vs. tape-travelling time

N^+ and Ti^+ are attracted by the dislocations and diffuse to the deep locations along the tube of the dislocation core. Therefore a new wear resistance layer is continually formed.

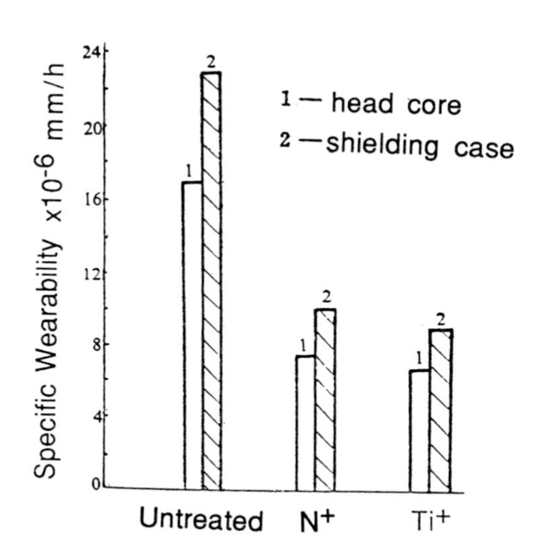

Figure 4 Specific wearability of head before and after implantation

The head core and the shielding case are made from different materials. Their hardness differs greatly before ion implantation and their wear depth also differs. Wear is severely inhomogeneous. After ion implantation the hardness of the shielding case increases and the wear depth decreases greatly. The difference in specific wearability between the core and the case is not large. The wear inhomogeneity is improved.

After ion implantation, the wear resistance of the head increases and the wear inhomogeneity improves. These are beneficial in improving the electro-magnetic parameters of the audio head.

The Electro-magnetic Parameter. Figure 5 illustrates the variation of the electro-magnetic parameters versus the tape-travelling time before and after N^+ implantation. After ion implantation, the variation in electro-magnetic parameters decreases.

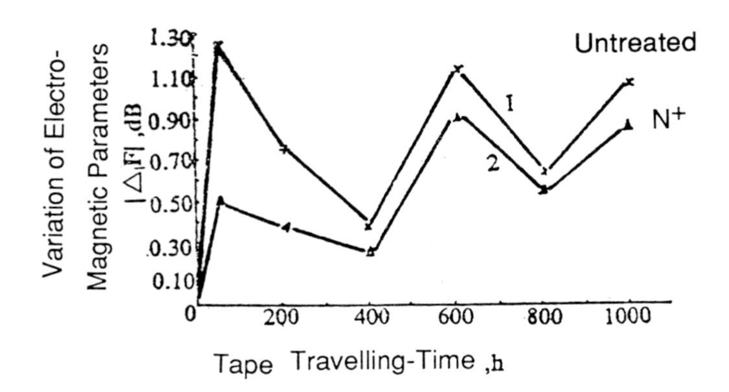

Figure 5 Variation of electro-magnetic parameters versus tape-travelling time

The variation in electro-magnetic parameters mainly depends on the head slot. When the tape is moving on the surface of the head, the depth of the slot decreases with increasing wear depth and the surface asperities of the head and the edge of the slot become rough, therefore the head distance loss increases and the electro-magnetic parameters become poor. After ion implantation the wear resistance increases. The wear depth decreases and the surface inhomogeneity improves, therefore the variation in electro-magnetic parameters decreases. Ion implantation is beneficial in prolonging the life-span of the head.

The Wear Appearance of the Audio Head

The wear appearance of the audio head is shown in Figure 6, and is the same before and after ion implantation. The main wear characteristics are scratches

Figure 6 Wear appearance of head

and pits. The scratches are distributed parallel to the tape travelling direction and the pits are scattered and rare. The scratches and pits are caused by Fe_2O_3 hard magnetic particles dropped from the audio tape moving against the surface of the audio head. The numbers of scratches and pits increase gradually with the increase in the tape-travelling time. The head wear is mainly due to micro-cutting by the magnetic particles. The head wear mode-identified as low stress scratching-is abrasive wear.

4 CONCLUSION

1. After a certain number of N^+ or Ti^+ ions are implanted on the surface of the permalloy audio head, its electro-magnetic parameters do not basically change.
2. After N^+ or Ti^+ implantation the hardness of the head increases and the wear resistance increases by over 100%. The wear inhomogeneity decreases. These are beneficial to improving the electro-magnetic parameters of the head and prolong the life-span of the audio head.
3. The thickness of wear resistance layer after N^+ or Ti^+ implantation is of the order of microns.
4. Wear of the audio head is characterized by scratches and pits. The wear mechanism is mainly by micro-cutting. The wear mode is abrasive-low stress scratching-wear.

REFERENCES

1. Y. Tu Materials of Mech. Eng., 1980, 2, 30.
2. Z. Xie, Materials of Mech. Eng., 1981, 5, 20.
3. H. Li, 'Ion Implantation and Change of Materials Surface Characteristics,' Science Press, Beijing, 1986.
4. F. Cui, Materials of Mech. Eng., 1981, 5, 1.
5. The State Standard of P.R. China, the General Technological Conditions of Audio Head of Cassette Recorders.
6. Y. Zhao, Beijing Electronic Sound, 1984, 1, 19.
7. N.E.W. Hartley, Thin Solid Films, 1979, 64, 177.
8. G. Dearnly and N.E.W. Hartley, Thin Solid Films, 1978, 54, 215.

2.1.6
Friction and Wear Between CVD TiC–TiN Composite Coating and Lead-base Brush Electroplating Coatings

R. Zhang

DEPARTMENT OF MECHANICAL ENGINEERING, TSINGHUA UNIVERSITY, BEIJING, 100084, PEOPLE'S REPUBLIC OF CHINA

1 INTRODUCTION

Many surface techniques have been used in several tribological fields. Solid lubrication coatings must be applied in extreme working conditions e.g. in space. Lead is one of the softest metals which also has good solid lubrication properties[1-3]. The lead-base solid lubrication films (for example, binary films of Pb-Sn and Pb-MoS_2 and ternary films of Pb-Sn-Cu, Pb-Sn-Ni and Pb-Sn-Co) are termed, "soft metal composite films". They have better lubrication properties than a pure Pb film because of the difference of mechanical and physical properties between lead and the other elements. Brush electroplating (BEP) is a conventional surface technique. When a hard surface slides against a solid lubricated surface, not only damage and failure of the soft surface layer must be considered, but also damage and failure of the hard surface layer must be investigated at the same time. The former could be retrievable under certain performance conditions, but the damage of the latter is always irreversible and disastrous. Therefore studies must be done for every rubbing pair composed of hard/soft surfaces. The CVD TiC-TiN composite multilayer coating is one of the best hard coatings. Each has a high melting point and hardness, high bonding strength with steel substrates, and low friction coefficients with other metallic materials[4-6]. The tribological properties of a CVD TiC-TiN composite coating ball sliding against BEP Pb-base coating disk have been studied in this paper. The results proved that this rubbing pair has good tribological characteristics, with many potential applications, for example, in astronautic sliding bearings.

2 SPECIMEN PREPARATION AND PROPERTIES

CVD TiC-TiN Composite Coatings

The coated ball specimen was prepared using low-pressure CVD equipment from the Bernex Corporation, Switzerland[5], at Jingwei Textile Machinery Factory, Shanxi, China. The whole experimental programme was controlled by microcomputer. In the deposition process, alternate layers of TiC and TiN constituted the multilayer composite coating. The reactions relevant to deposition are:

$$TiCl_4(g) + CH_4(g) \underset{950-1050^0C}{\overset{H_2}{\rightleftharpoons}} TiC(s) + 4HCl(g) \quad (1)$$

$$TiCl_4(g) + \tfrac{1}{2}N_2(g) + 2H_2(g) \underset{850-1000^0C}{\overset{H_2}{\rightleftharpoons}} TiN(s) + 4HCl(g) \quad (2)$$

$$TiCl_4(g) + CH_4(g) + \tfrac{1}{2}N_2(g) \underset{900-1050^0C}{\overset{H_2}{\rightleftharpoons}} Ti(CN)(s) + 4HCl(g) \quad (3)$$

The deposition time of the composite coating and the thickness of each layer are shown in Table 1. The main technical parameters, such as the deposition temperature at each stage, the concentration ratio of the mixed gases, the velocity of the gas flow, the deposition time and the selection of the speed of rotation of parts, were stored

Table 1 Thickness and deposition time for each layer in CVD TiC-TiN coating (deposition temperature: 1000^0C)

Deposition order	TiC	TiC_xN_y	TiC	TiC_xN_y	TiC	TiC_xN_y	TiN	total
Thickness (μm)	1	0.5	1	0.5	1	0.5	1	5.5
Deposition time (h)	1	0.5	1	0.5	1	0.5	1	5.5
Lattice const. (nm)	0.429	0.425	0.427	0.425	0.427	0.425	0.4235	

on tape. Thus the accuracy, reliability and repeatability of the program controlling the process could be guaranteed.

Brush Electroplating (BEP) Pb-base Coatings

AISI 1045 steel substrates were brush electroplated with pure lead or Pb-Sn-Cu. Disks of the substrate, quenched at 860^0C, had external dimensions of 5mm thickness and 63.5mm diameter, the surface roughness was in the range 0.15-0.2μm CLA. The input voltage of BEP equipment model 2KD-I (made in China) was 220 V; the DC output voltage 0-30 V; the output current 0-30 A, the usual power consumption coefficient for brush plating C_{bp}=0.101 A h mm^{-2} μm^{-1}. All of the solutions for BEP coatings were provided by the Department of Mechanical Engineering, Tsinghua University, Beijing: including electroactivation solutions (Nos. 2 and 3), electrocleaning solution, special Ni BEP solution (for the base-layer) and pure Pb and Pb-Sn-Cu BEP solutions. Before plating, the specimen was ultrasonically cleaned using acetone and then ethyl solutions. The main parameters for BEP preparation are listed in Table 2. After every step the specimen was washed in cold water. The ambient temperature during the plating process was 20-22^0C with a relative humidity of 50%. The BEP disks were put into a baking furnace at 50-55^0C for 30 minutes, then preserved in a dryer. By means of energy dispersive spectroscopy (EDS) analysis it was determined that the composition of L2D film was: Pb=76.4 wt%, Sn=12.6 wt%, Cu=11.0 wt%, the L1D was almost pure Pb.

Specimen Preparation

The contacting couple was a fixed ball and rotating disk. The ball was 7.14 mm in diameter, the disk was 63.5 mm in diameter and 5 mm thick. The balls were of two types: AISI 52100 steel uncoated (SB) and CVD TiC-TiN composite coated (CB). The CVD coated balls were oil-quenched in vacuum after deposition. There were also two kinds of disks: BEP pure Pb (L1D) and Pb-Sn-Cu (L2D) coated on AISI 1045 quenched substrate. The roughness of the uncoated ball was 0.1 μm, but that of the CVD coated ball was less than 0.1 μm. The roughness of the steel

Table 2 The parameters for the brush electroplating process

	voltage V_a (V)	relative speed V_{ps} (m min^{-1})	process time l_{ps} (min)	thickness t (μm)
electrocleaning	12	10	0.67-1	
activation solution No. 2	12	15	0.15	
activation solution No. 3	12	15	0.33-0.67	
Ni base-layer	8	10	3	1.2-1.5
Pb coating	12	8	12	18.4
Pb-Sn-Cu coating	14	8	15	20.3

disk (substrate of the BEP coating) was 0.8 μm, but those of the BEP Pb and Pb-Sn-Cu coated disks were 0.25 μm and 0.20 μm respectively. The hardness of the specimens were: CVD coating, 2550-2650 HV_{50g}; BEP pure Pb coating 23-28 HV_{20g}; BEP Pb-Sn-Cu coating 34-40 HV_{20g}; AISI 1045 steel, 59 HRC; AISI 52100 steel, 62 HRC. The experiments were conducted with and without oil lubrication - a good vacuum oil lubricant made in China, model SP8801-100 (trihydroxy methyl-propyl ester mixture)[7], with a saturation pressure of 3.9×10^{-5} Pa, a viscosity of 100 cSt (40^0C) and a condensation point of about -50^0C.

3 EXPERIMENTAL METHODS AND RESULTS

Friction and Wear Tests

The friction and wear tests were carried out on a model MT-1 vacuum friction and wear testing machine designed and manufactured by ourselves[8]. The contact form was ball-on-disk; the vacuum was 6.67×10^{-3} Pa; the upper specimen (ball) was fixed; the lower specimen (disc) was rotating. The running-in distance was 18 m (at a sliding speed of V_0=0.3 m s^{-1}). The drop-feed mode was adopted for oil lubrication. Measurement started after reaching the boundary lubrication state. When the average sliding friction coefficient μ vs. sliding time t curves were measured, from sliding time t=0 to t=50 minutes each point of the friction coefficient μ represented the average value of five measurements; the sampling frequency **γ** was 5 s^{-1}. When the μ-load P (fixed V_0) and the μ-V_0 curves were measured the adopted value of μ was the average value of 10 minutes after the running-in period. The method for measuring the wear rate of the ball was based on observing the section size of the wear track from which the volume wear rate (cubic millimetres per metre) can be calculated.

The Variation of μ with t. Table 3 shows the value of μ with constant speed (V_0=0.5 m s^{-1}) and load (P=5 N) in dry friction and oil lubrication conditions respectively. The sliding time t_0 was 50 minutes. The rubbing pair was: SB-L1D, uncoated steel ball against BEP Pb coating disk; SB-L2D, uncoated steel ball against BEP Pb-Sn-Cu coating disk; CB-L1D, CVD TiC-TiN coated ball against BEP Pb coated disk; CB-L2D, CVD TiC-TiN coated ball against BEP Pb-Sn-Cu coated disk. In order to compare with the results of uncoated steel disks, the rubbing pairs of SB-SD (uncoated steel disk) and CB-SD have to be mentioned also. Their results can be seen in detail in reference 9. From these values we know the following:

1. In air without oil lubrication, the friction coefficient μ of CB-L1D is almost the same as that of CB-L2D, and μ increases up to about 0.6 when sliding time t>25 min. for CB-L2D and t>50 min. for CB-L1D. It is noted that the CVD TiC-TiN composite coating started to crack at t=25 min. and 50 min. respectively. But at the initial stage, the friction coefficient of CB-L2D is higher than that of CB-L1D. Both are much lower than those for the uncoated disk[9]. The value of μ for the coated ball is lower than the uncoated ball, but μ for the coated ball is higher than that for uncoated ball after the CVD TiC-TiN composite coating was broken.
2. In vacuum (6.67×10^{-3} Pa) without oil lubrication, μ for CB-L2D is higher than that of CB-L1D, which is always lower than 0.2 (the average value is 0.17) i.e. the lowest value in the four rubbing pairs (CB-L1D, SB-L1D, CB-L2D, SB-L2D). Both are also very much lower than those for the uncoated disk[9].
3. With oil lubrication, μ for the solid lubricated pair decreases further, the CB-L2D rubbing pairs are the

Table 3 Sliding friction coefficient μ (P=5 N, V_0=0.5 m s^{-1}, t_0=50 min.)

	Dry friction						Oil lubricant: SP8801-100					
	SB SD	SB L1D	SB L2D	CB SD	CB L1D	CB L2D	SB SD	SB L1D	SB L2D	CB SD	CB L1D	CB L2D
Vacuum	.47	.32	.28	.26	.17	.18	.053	.052	.041	.052	.043	.032
Air	.68	.40	.37	.46*	.32*	.26**	.086	.079	.062	.081	.073	.050

*After about 50 min. CVD coating wore out, then μ=0.61
**After about 25 min. CVD coating wore out, then μ=0.63

best in both air and vacuum. The lowest value of μ, about 0.03, is obtained in vacuum for the CB-L2D rubbing pair.

Variation of μ with P and V_0. The μ-V_0 (P=5 N) and μ-P (V_0=0.5 m s^{-1}) relationships are obtained for four rubbing pairs (CB-L1D, SB-L1D, CB-L2D, SB-L2D) with oil lubrication (Figure 1). The following can be observed:

1. The curves of μ-V_0 usually decrease with an increase of V_0, but the curves of μ-P have a maximum value usually at P=18 N. Also the slope of this kind of curve is steep at smaller P, then the variation of μ becomes very smooth, for large loads, μ decreases slightly, i.e. they have the same shape of curve for each group.
2. At P=3 N, the μ of CB-L1D is lower than that of SB-L1D, then the difference becomes smaller. The μ for Pb-Sn-Cu BEP disk has similar characteristics.
3. For CB-L1D and SB-L1D rubbing pairs, the difference between their μ-V_0 curves is small, especially in a vacuum. But it can be seen from the μ-V_0 curves for CB-L2D and SB-L2D rubbing pairs that the CB-L2D rubbing pair very much improves the tribological properties.

Volume Wear Rate W_V. Table 4 shows the volume wear rate for the four rubbing pairs; we can deduce the following:

1. The value of W_V of the ball is much smaller with an oil lubricant than without oil lubricant.
2. The value of W_V of the coated ball is much smaller than that of uncoated steel ball. If the coated ball

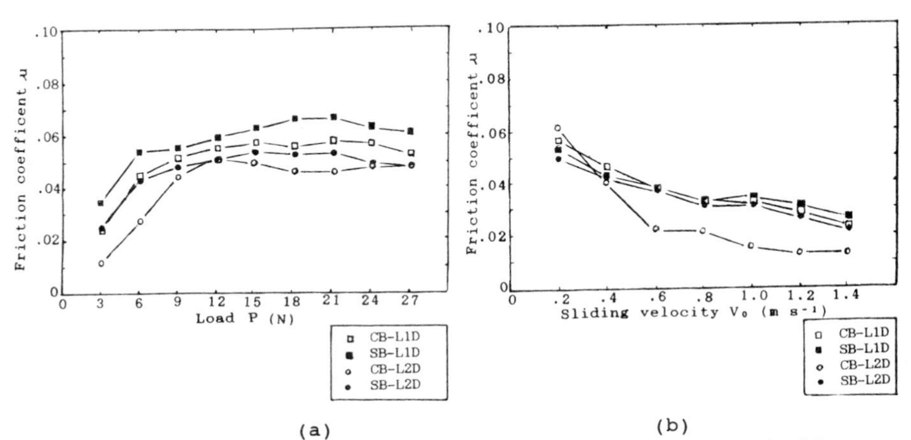

Figure 1 μ-P and μ-V_0 curves for steel and coated balls on Pb and Pb-Sn-Cu BEP coated disks with oil lubricant in vacuum

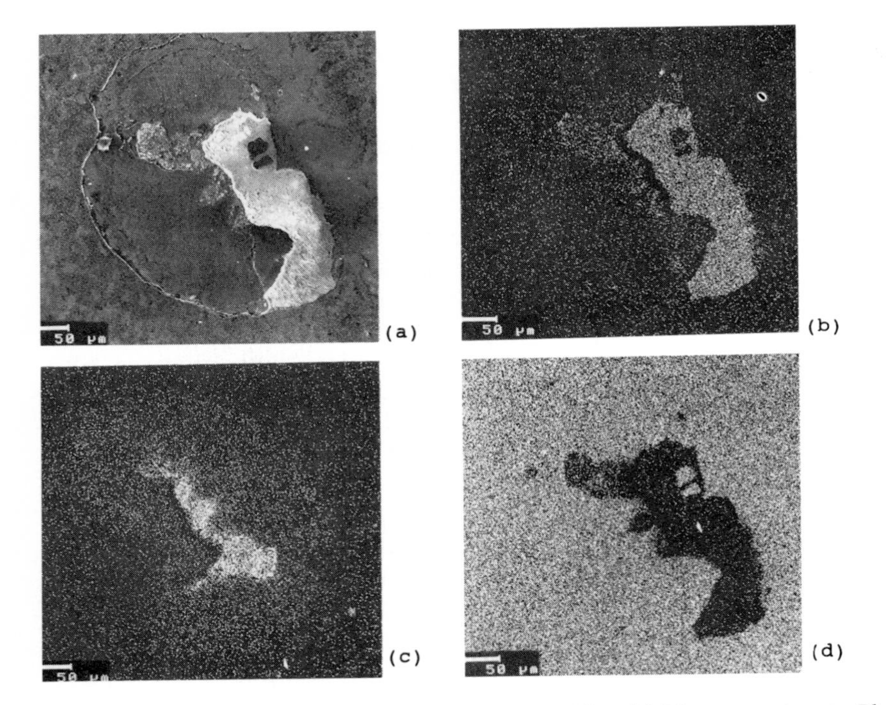

Figure 2 Wear track of CVD coated ball sliding against Pb-Sn-Cu BEP coated disk and its X-ray mappings (P=5 N, V_0=0.5 m s^{-1}, t=50 min., in vacuum, oil lubricant SP8801-100):
(a) SEM micrograph (b) Pb X-ray map
(c) Fe X-ray map (d) Ti X-ray map

is nearly worn out, the difference becomes small.

3. The most important factor is that the two kinds of balls have very different wear properties in a vacuum. For the SB-SD pair, the tendency to adhesion increases greatly in a vacuum so that the value of W_V of the steel ball is about one order of magnitude larger than that of the coated ball. If the disk is covered by a Pb-base BEP layer, the value of W_V decreases significantly.

SEM Observation and AES Composition Profile Analysis

The wear tracks of the balls and disks were observed using a CSM 950 scanning electron microscope (SEM) and analysed with a TN 5402 Series II energy dispersive spectrometer (EDS) linked to the SEM; the compositional profile analysis for the micro regions of the worn surfaces of specimens was obtained by PHI 600 multi-function Auger electron spectroscopy (AES).

SEM Observation of Wear Track on the CVD Ball. Figures 2 and 3 show the wear tracks of balls (CB and SB) sliding against the disk (L2D) in the oil-lubricated condition (P=5 N, V_0=0.5 m s^{-1}) in vacuum (6.67x10^{-3} Pa) respectively. They show that:

1. the CVD TiC-TiN coating was partly worn in the contact area;
2. the transfer solid lubrication film was formed only at the back of the worn region on CVD coatings and not on the whole worn region;
3. the transfer solid lubrication film was also formed on the worn region of the SB ball, but it was not continuous;
4. the X-ray distribution maps for elemental Fe were also different for the two wear tracks, there is no elemental Fe in the transfer film on the CVD coated ball.

SEM Observation of Wear Track on the BEP Disk. Figure 4 illustrates the wear tracks on the Pb-Sn-Cu BEP disk with oil-lubrication in (a) vacuum and (b) air. They show that:

Table 4 Volume wear rate W_v of balls ($\times 10^{-6}$ mm^3 m^{-1})
($P=5$ N, $V_0=0.5$ m s^{-1}, $t_0=50$ min.)

	Dry friction						Oil lubricant: SP8801-100					
	SB SD	SB L1D	SB L2D	CB SD	CB L1D	CB L2D	SB SD	SB L1D	SB L2D	CB SD	CB L1D	CB L2D
Vacuum	13.6	1.9	1.7	.80	.83	.81	.89	.28	.10	.020	.041	.066
Air	1.6	4.4	5.8	1.13*	1.1*	1.3**	.075	.21	.17	.060	.014	.057

* After about 50 min., CVD coating wore out.
** After about 25 min., CVD coating wore out.

1. the L2D solid lubrication film was worn out for every experimental condition, but the widths of the wear track were different;
2. the wear track in air was wider and deeper than that in vacuum, indicating that wear was more severe in air than in vacuum.

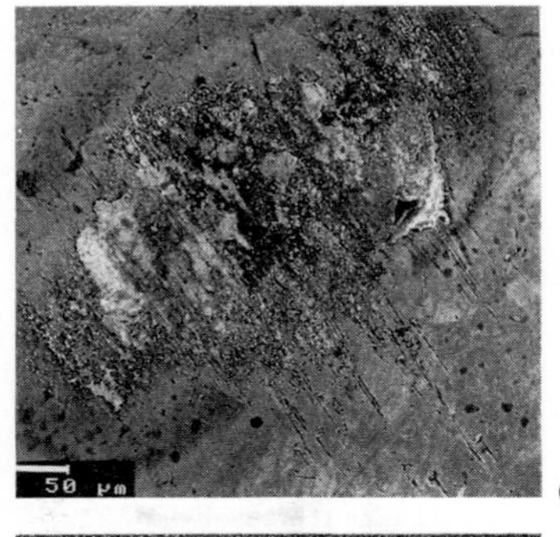

Figure 3 Wear track of CVD coated ball sliding against Pb-Sn-Cu BEP coated disk ($P=5$ N, $V_0=0.5$ m s^{-1}, $t=50$ min., in vacuum, oil lubricant SP8801-100)
(a) SEM micrograph
(b) Pb X-ray map
(c) Fe X-ray map

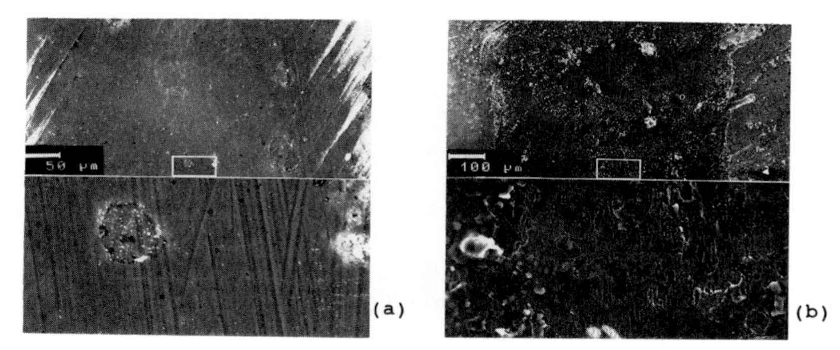

Figure 4 Wear tracks on Pb-Sn-Cu BEP coated disk matched with CVD coated balls with oil lubrication (P=5 N, V_0=0.5 m s^{-1}, t=50 min.):
(a) in vacuum (b) in air

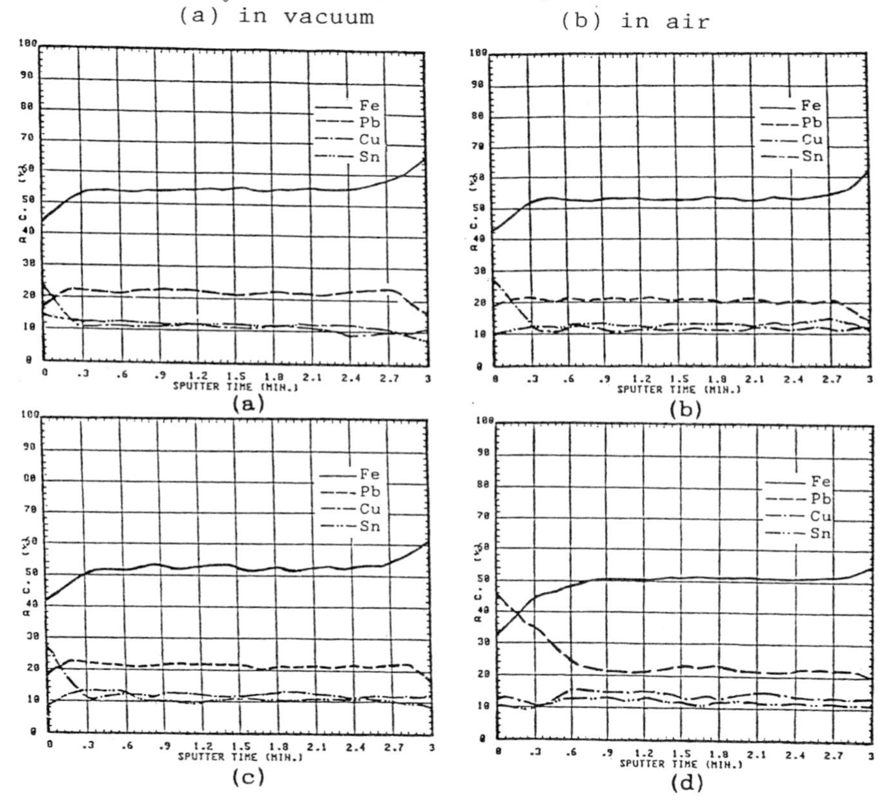

Figure 5 AES composition profile curves for wear track on Pb-Sn-Cu BEP coated disk matched with CVD coated balls with oil lubrication (P=5 N, V_0=0.5 m s^{-1}, t=50 min.)
(a) at centre of the track (D=0) (b) D=0.1 mm
(c) D=0.2 mm (d) D=0.3 mm

AES Composition Profile Analysis of the BEP Disk. AES surface composition analysis was done for the wear tracks on Pb-Sn-Cu BEP disks, after sliding against the CVD TiC-TiN composite coated ball. Figure 5 is the AES composition profile of the same specimens with oil lubrication in vacuum. D is the distance from the centre of the wear track on the worn Pb-base BEP disk. Clearly:

1. There is always a thin solid lubrication film on top of the wear track of the Pb-base BEP disk, even without oil lubrication. So there is always the lubrication action which decreases the sliding friction coefficient μ and the volume wear rate W_v.
2. There is Fe in the thin film, the composition of which is different to that of the Pb-base BEP coating. Evidently, this thin film is not simply from the original coating, but the result of mutual action between friction surfaces.
3. The thickness of the thin film is smallest when Fe content is highest at the centre of the wear track; as D increases, the thickness increases and the Fe content decreases gradually, but the concentration ratios of Pb, Sn and Cu remained roughly the same.

4 CONCLUSIONS

1. The experimental results on the friction coefficient and volume wear rate show that the Pb-base solid lubrication coatings have better tribological properties in vacuum than in air.

2. Contributions from Sn and Cu in Pb-Sn-Cu BEP coating, enhance tribological properties over those of pure Pb BEP coating.

3. The solid lubrication transfer film forms on top of the wear track of both ball and disk, and is the result of mutual action between friction surfaces.

REFERENCES

1. M.J. Todd and R.H. Rentall, 'Lead Film Lubrication in Vacuum,' Proc. ASLE Conf., Denver, CO, 1978.
2. J. Gerkema, Wear, 1985, 102, 241.
3. C.Y. Shih and D.A. Rigney, Wear, 1989, 134, 165.
4. H.E. Hintermann, Tribol. Int., 1980, 13, 267.

5. Bernex Company, 'Moderate-temperature CVD Brochure,' Berna AG 01-ten, Switzerland, 1982.
6. R. Zhang, Z. Lin, Z. Cui and Q. Song, Wear, 1991, 147, 227.
7. Z. Cui, Q. Song, R. Zhang and Y. Fan, J. Beijing Manag. Inst. Mach. Build. Ind., 1989, 4:1, 108, (in Chinese).
8. R.Zhang and Z. Cui, Physics, 1988, 17, 681, (in Chinese).

2.1.7
Wear Resistant Coating Compositions Based on Modified Phenol Resins

A.L. Zaitsev

METAL–POLYMER RESEARCH INSTITUTE, ACADEMY OF SCIENCES, GOMEL 246652, BELARUS

1 INTRODUCTION

Friction characteristics of solids under boundary lubrication conditions are known to depend significantly on the adsorption processes occurring during contact; in turn, these processes depend on the physical and chemical structures of the contacting surfaces[1]. Data are available indicating that the nature of the solid surface influences the physical state of the adsorbed hydrocarbon molecules which affect in a particular manner the friction properties of the lubricating layer[2]. Hence, friction properties under boundary lubrication conditions can be successfully controlled by using composites of known compositions; this allows a continuous lubricating film to be formed on the surface thus protecting the surfaces in contact against direct interaction under severe friction conditions. Polymer-based composites suit most nearly the above situations because of excellent processability and wide ranges of properties; therefore, it was of interest to investigate friction properties and frictional interaction of composite coatings lubricated with mineral oil, and also to understand the effect of the filler chemical structure.

Since friction behaviour of thermoplastic polymers under three conditions of liquid lubrication has been understood quite adequately, materials based on phenol resin about which little information is available were the subject of this investigation.

2 EXPERIMENTAL

Resol phenol-formaldehyde resin, organic and mineral fillers were studied. During processing of synthetic and natural fibre, rejects are usually available. Attempts have been made to utilise them in reinforcements for composite coatings. Poly(acrylonitrile) (PAN), polyamide (PA), polyoxadiazole (PODA), rayon, cotton, poly-(ethylene terephthalate) (PETP), asbestos and glass fibre were investigated. Antifriction fillers were polytetrafluoroethylene (PTFE), molybdenum disulphide (MoS_2) and partially graphitized carbon. Dispersed iron, iron oxide, aluminium oxide, titanium oxide and titanium carbide (Table 1) were used as the structural materials.

Friction tests were performed on a machine where two hollow cylinders rubbed faces at pressures from 0.2 to 1.0 MPa and sliding velocity of 2.5 m/s. Tests lasted from two to seven hours. The friction force, temperature of oil and linear wear rate were registered. Wear volumes of the steel counterface were determined by weighing using an analytical balance, while linear dimensions of the coatings altered in the course of testing were measured by a micrometer.

Steel 3 was tested lubricated with engine oil $M8B_1$ (USSR Standard GOST 10571-88) containing 4-7% additives of calcium, phosphorus and zinc compounds. Oil, 10 ml, was poured into a special mandrel and maintained there during the entire testing period. The effectiveness of fillers was estimated by comparison with phenol resin and high grade iron.

3 RESULTS

The wear rate of the coatings in question was found to be dependant on the filler origin. The wear resistance of the coatings measured at higher loads was lower for fibre fillers than for particulate ones, the exception being cotton and rayon fibre. No wear was registered for the unfilled binder because of poor coating adhesion; the oil did not penetrate the adhesional seam which resulted in coating tear-away during testing. When compared to iron-against-steel characteristics it can be seen (Figure 1) that only a few fillers can compete with iron in wear resistance, e.g. cotton, rayon, ferric oxides, aluminium oxides and iron powder. Some fillers (MoS_2, cotton, Al_2O_3, PODA) form somewhat thicker coatings.

Table 1 Chemical structure and concentration of filler in phenol-formaldehyde-based coatings

Fiber filler				Particulate filler			
Filler	Chemical structure	%	Micro-hardness, MPa	Filler	Chemical structure	%	Micro-hardness, MPa
PA	$[-(CH_2)_x-CO-NH-]_n$	50	297				
PODA	$[-C_6H_4-C(=N-N=)C(-O-)-]_n$	50	357	PFR	$[-C_6H_2(OH)-CH_2-]_n-OH$	100	620
PAN	$[-CH_2-CH(CN)-]_n$	50	325	PTFE	$(CF_2 - CF_2)_n$	60	90
					MoS_2	40	600
Cotton	$[-O-CH(CH_2OH)-CH(OH)-CH(OH)-CH-O-CH-]_n + [C_6H_3(OH)(O-CH_3)(CH=CH-CH_2OH)]_m$	58	437	Graphitic carbon	Amorphous and crystalline carbons	40	460
					Fe_2O_3	54	770
Rayon	$[-O-CH(CH_2OH)-CH(OH)-CH(OH)-CH-O-CH-]_n$	51	690	Iron, <1 μm	Fe	70	1880
					Al_2O_3	51	1770
PETP	$[-(CH_2)_x-O-CO-C_6H_4-CO-O-]_n$	43	370				
					TiO_2	50	1170
Asbestos	$2SiO_2 \cdot 2H_2O \cdot 3MgO$	54	1240		TiC	50	350
Glass	Na_2O, $CaO \cdot 6SiO_2$	48	790				

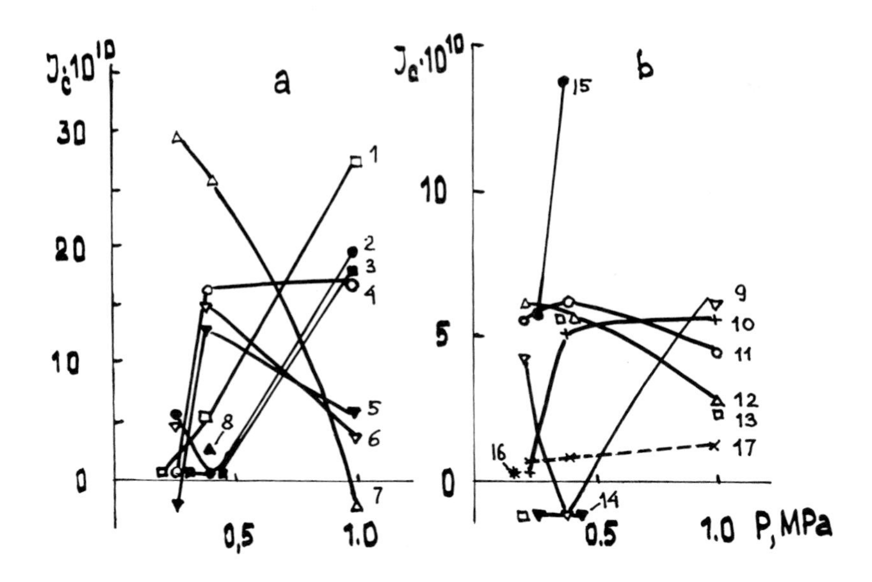

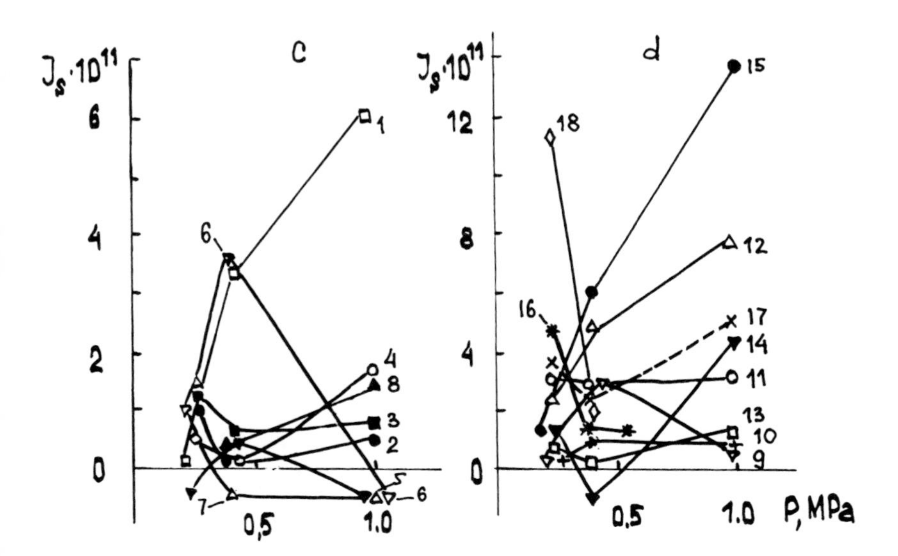

Figure 1 Wear rate of coatings (a,b) and counterface (c,d) vs. load and filler origin when lubricated with mineral oil: (a,c) fibres; (b,d) powder.
1. asbestos; 2. PAN; 3. glass; 4. PA; 5. cotton; 6. rayon; 7. PODA; 8. PETP; 9. TiC; 10. PTFE; 11. Fe_2O_3; 12. Fe; 13. Al_2O_3; 14. MoS_2; 15. TiO_2; 16. phenol; 17. steel-iron; 18. graphitic carbon

The wear resistance of steel counterfaces was observed to vary depending on the steel origin. Steel was measured to have the highest wear resistance if organic fibre, Al_2O_3, PTFE and Fe_2O_3 powders were the fillers. It should be noted that when rubbed against a composite containing cotton, rayon or PODA fibre, and also MoS_2 the steel surface underwent heavy oxidation leading to significant weight increments.

The analysis of the friction coefficient vs. load dependence indicated that the friction coefficient variation mode is determined by the filler used; on the pressure rising it could increase or decrease, Figure 2. The lowest friction coefficient was measured for the phenol polymer filled with PTFE, PETP, cotton and Fe_2O_3. The other gave higher friction coefficients than the steel-iron pair.

The oil temperature was seen to increase in a monotonous manner with increasing load for the fibre fillers while for several particulate fillers the oil temperature fell after the contact pressure increased. Comparison of the oil temperature and the friction coefficient showed that little heat was released with PTFE and Fe_2O_3 fillers possessing low friction coefficients.

4 DISCUSSION

The above results indicate the filler nature and how its physical and chemical structure influence the boundary friction characteristics. The fillers selected for the investigations are rather difficult to classify by their mechanical properties, since the latter affect the frictional behaviour in the initial period of operation. After the run-in period has been completed, a layer of adsorbed lubricant, tribochemical interaction products from the counterface, filler, binder and mineral oil was found on the coating. This layer controls lubrication and friction behaviour of the coating within the steady friction regime.

The filler chemical structure was found to be responsible for a wide scatter in wear resistance data ($0.1 - 25x10^{-10}$) which was believed to be related to the altered oil-filler adsorptive interaction. The increased or decreased coating thickness measured at low loads

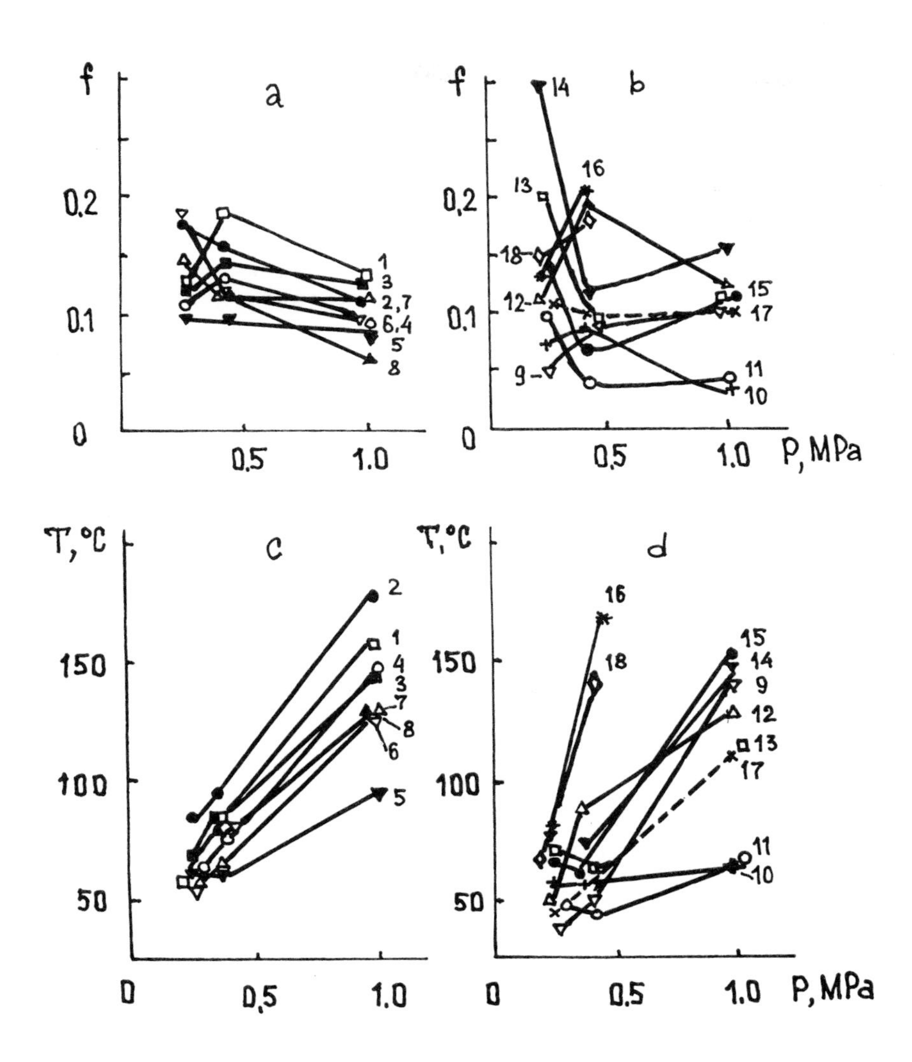

Figure 2 Friction coefficients of coatings (a,b) and oil temperature (c,d) vs. load and filler origin - designations as in Figure 1

resulting, respectively, from the volume oil absorption and surface layer oil absorption, and from extraction of low-molecular weight components present in the coating is evidence of areas degraded by liquid lubricants. In composites such areas may be interphase interfaces rich in reaction products between the filler and the binder.

The microhardness measurements performed on the rubbed coating surfaces indicated that most fillers retained oil in their surface layers (exceptions being PTFE, PETP, cotton, rayon and asbestos fibre, TiC, Fe and MoS_2). Among those retaining oil, metal oxides and nitrogen-containing organic substances (PA, PAN and PODA) are found.

The analysis of filler chemical structures which caused severe oxidation of steel during rubbing showed the molecules of cotton, rayon and PODA fibre to incorporate reactive oxygen-containing (e.g. ether, methylol and hydroxyl) groups. Probably, in the hydrocarbon environment steel oxidizes under the effect of oxygen-containing molecules of the filler. The positive result of a given filler is that owing to filler degradation and tribochemical reactions, oxide films are formed on steel possessing high adsorption power towards oil. This is reflected in a relatively low friction coefficient of coatings reinforced with rayon and cotton fibre.

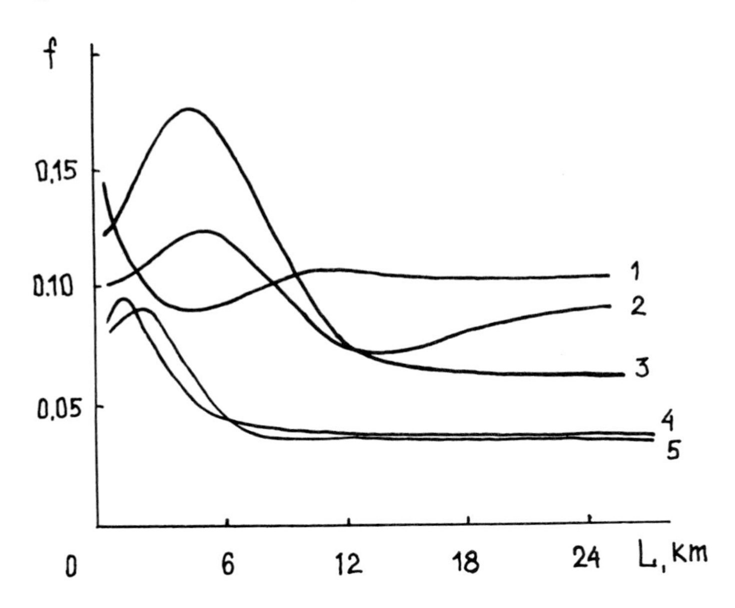

Figure 3 Friction coefficients of modified phenol coatings vs. operation of steel in pairs with: 1. iron, 2. PA and PFR; 3. PA and PFR modified with epoxy oligomer; 4,5. composition based on modified phenol containing polymeric additives and mineral filler

The specific features of frictional interaction allowed the development of composite coatings capable of strong adhesion to steel and being highly resistant to oils, water and benzine. The binder was phenolic resin-modified with an epoxy oligomer. The number of reactive oxygen-containing groups in the binder increased thus leading to lower friction coefficients and wear when the coatings were operated at the boundary conditions. Figure 3 illustrates the dependencies of friction coefficients vs. time for coatings containing the modified binder, and also for several compositions prepared with it. It can be seen that the compositions are characterized by low friction coefficients as compared with the base variant, i.e. steel - iron pair.

5 CONCLUSION

The results provide evidence that polymeric composite coatings based on phenolic resins can be used in place of metals in friction units of machines operated under the conditions of limited liquid lubrication, and also for repair purposes and recovery of worn elements.

REFERENCES

1. S. R. Morrison, 'The Chemical Physics of Surfaces,' Plenum Press, New York and London, 1977 (Russian translation).
2. V. V. Smurugov, and I. D. Delikatnaya, 'Effect of Polymer Surface on Oil Boundary Layer Structure,' Soviet Journal of Friction and Wear, 1988, 9,(4), p.739.

2.1.8
Increase of Polymer Coating Wear Resistance by Electrophysical Modification

V.S. Mironov

METAL–POLYMER RESEARCH INSTITUTE, ACADEMY OF SCIENCES, GOMEL 246652, BELARUS

1 INTRODUCTION

Frictional interaction of polymers and metals is known to be followed by intensive electrophysical processes, e.g. electrification, electron emission and electrical polarization. These influence significantly the structure and triboengineering characteristics of polymer materials[1-3]. It is believed that antifrictional capacity and wear resistance of tribological polymeric elements can be increased by controlling these processes using various means.

Bilik[3] and Savkoor[4] reported lower levels of electrification and suppression of triboelectricity for metal-polymer tribosystems by charge neutralization through the external electric field superposition to give lower friction coefficients and higher wear resistance of the contacting materials.

The above effects may be interpreted by taking into consideration the energy balance of friction described by Fleischer[5], as related to the decreased portion of work of the friction force required to form the triboelectret state in the polymeric friction element. With this in view, of particular interest is to achieve an understanding of the effect by which the electret state in polymeric frictional elements can be attained by physical means.

The purpose of the present work was to study the influence of thermal treatment in an electric field and treatment by a beam of non-penetrating electrons - these methods appear the most efficient in preparing polymer electrets[6] on the wear resistance and physicomechanical properties of coatings based on thermoplastic polymers.

2 EXPERIMENTAL

The following coatings were used in the experiments: high-density polyethylene (HDPE Grade 21006-075, GOST 16338-77), pentaplast (PP1 poly-/3, 3-bis-(chloromethyl) oxetane/, Grade P-A-2, TU 6-05-1422-79), polychlorotrifluoroethylene (PCTFE GOST 13744-76) and polycaproamide (PCA Grade PA6-120/321, OST 6-06-09-76). These materials were selected owing to their application in triboengineering, and also to their different polarities, temperatures of phase transitions and irreversible radiation effects.

Coatings of 0.36 to 0.42 mm thickness were moulded from molten polymers on aluminium foil. Methods used to prepare coatings, to study the physicomechanical properties - tensile strength and microhardness - and structures of the polymer materials were similar to those described in References 2 and 7.

The coatings were subjected to thermal electretization in constant electric fields of intensity $E = 0.5$ to 60 MV/m at temperature $T = 380$ to 390 K; the exposure time was 1.8 ks; they were subsequently cooled in the field at the rate of ca. 0.2 K/s. Dielectric teflon layers were used during the thermal electretization.

Polymer surfaces were electron treated in vacuum of 6.7×10^{-3} Pa with a beam of electrons having a monoenergy of 25, 50 and 70 keV, and current density of 10.2 mA/m^2 following the procedure reported in Reference 7. The effective surface charge density (ESCD) σ_{eff} was estimated by a non-contact compensation technique using an oscillating electrode[6].

The coatings were friction tested in pair with stainless steel counterfaces in air. The geometry shaft-sector bearing (K=1/6) was used; the sliding velocity V = 0.5 m/s and nominal pressure p = 0.1 - 0.5 MPa were assumed as described elsewhere[2,7].

3 RESULTS AND DISCUSSION

Before proceeding with the effects of thermal electretization and electron treatment, it seems useful to briefly state the main mechanisms by which electrification influences the friction characteristics of polymer materials.

Effects of Electrification

The study of the rôle and singularities of electrification during polymer friction allowed three main mechanisms to be distinguished through which electrification can affect the friction parameters of polymer-metal tribosystems: 1) generation of pondermotor forces between the contacting surfaces, being the motive force of mass transfer in the friction zone and leading to higher adhesion (pressure) in the contact; 2) variation in the properties of the polymer surface layer, and 3) creation of triboelectret state and variation of structure in the volume. The latter mechanism has been much neglected by researchers.

It was noticed that mutual electrostatic repulsion of trapping centres (of surface and bulk states[6]) filled with tribocharge carriers preferably localized in the actual contact spot region cause - in the polymer surface layer - tangential tensile stresses. The latter weaken the dispersion force of the polymer surface and increase wear. The electrostatic interaction of the charged areas and metal substrates gives rise to the normal compressive stresses which tend to compact the polymer material, flatten the roughnesses and decrease wear of the polymer.

The appearance of charged patches should increase the adsorptive activity of the polymer surface (better wettability, thermal oxidation, etc). The joint effect of internal electric field of the surface charge whose intensity can reach 10 MV/m or more - for the case of thin films - and the heat generated in the friction zone causes polarization of the bulk polymer material leading to the triboelectret state.

The residual polarization is largely related to the orientation of the polar groups and segments of the macromolecules perpendicular to the friction surface under the field effect. The orientation of the molecular structure in the course of forming the triboelectret state is, in our opinion, one of the causes that can decrease and stabilize friction forces and wear rate of the polymer element in a tribosystem after the run-in period. This agrees well with the known data on improved wear resistance of polymer materials and composites, having the structures oriented to the friction surface[8].

Effect of Thermal Treatment in an Electrical Field

The investigations conducted showed thermal electretization to cause significant variations in the wear resistance and physicomechanical properties of the polymer coatings tested. Figures 1 and 2 depict relationships between the relative tensile strength σ_p^E/σ_p and the relative wear rate J^E/J (σ_p^E, J^E and σ_p, J are strengths and wear rates of coatings treated in field E_e and untreated) and polarizing field intensity at constant temperature and time parameters of thermal electretization.

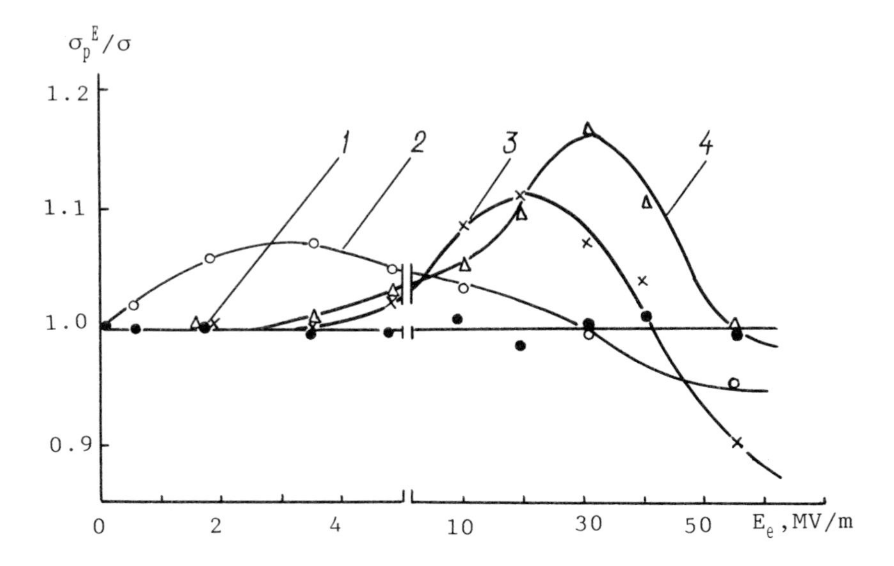

Figure 1 Dependence of relative tensile strength of coatings: HDPE (1), PP1 (2), PCTFE (3) and PCA (4) on the intensity of electric field during thermal treatment

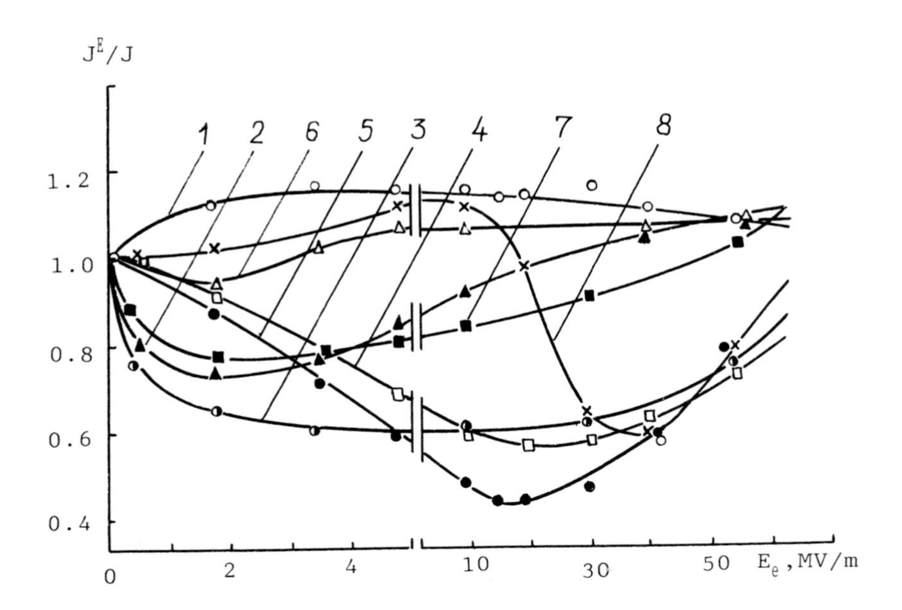

Figure 2 Relative wear rate after one hour (1-4) and subsequent six-eight hours of sliding (5-8) vs. intensity of electric field during thermal treatment of coatings: HDPE (1,5), PP1 (2,6), PCTFE (3,7) and PCA (4,8)

It should be emphasized that a similar thermal treatment of coatings without application of field E_e would not lead to meaningful variations in the friction and strength properties of the coatings. Examination of the relationships in Figure 1 indicates the coatings of polar polymers - PP1, PCTFE and PCA to have their strengths increasing in an extreme mode with increasing field intensity E_e. The coatings of non-polar polymer (HDPE) would not in fact change their strengths.

As the result of thermal electretization the deformation properties of the coatings deteriorated. X-ray structural analysis and IR spectroscopy revealed partial recrystallization, which can increase by 5-7% the crystallinity of the polymer coating material.

The simultaneous treatment of the mechanical test data, those of thermal depolarization and IR-spectroscopy led to a conclusion that higher strengths of the coatings under consideration are related to the orientation of macromolecular segments, ordering of the structure, and effect of the volume charge field on the process of dipole-segmental motion in the polymer. With increasing the polarity and glass-transition temperature of the polymer material the effect of strengthening intensifies and shifts towards the higher values of the electric field intensity E_e.

The relationships in Figure 2 $J^E/J = f(E_e)$ obtained from the data of the first sliding hour (curves 1-4) and subsequent 6-8 hours (curves 5-8) are quite different. For the case of polar polymers - PP1, PCTFE, and PCA - thermal electretization decreases considerably the value of J during the run-in period (curves 2-4), while for non-polar HDPE, this was observed during the steady friction period (curve 5). Unlike the polar polymers, HDPE exhibited higher J values during the run in period owing to thermal electretization. Comparison of respective relationships ($J^E/J = f(E_e)$ and $\sigma^E_p/\sigma_p = f(E_e)$) shows that the polar polymers have minimal wear rates after reaching maximum strength values. This fact witnesses for a close interrelation between structural changes and wear resistance of the polymer coatings and appears helpful in distinguishing the structural modification as one of the major aspects of the effect the treatment in an electric field causes - Aspect I.

Since thermal electretization leads to the electret state in polymer coatings whose main characteristic is σ_{eff}, it was important to obtain dependences of J on σ_{eff}, the sign and magnitude of which could be varied by changing the direction and magnitude of the field intensity.

Figure 3 shows the relationship $J^E/J = f(\sigma_{eff})$ constructed for HDPE and PCTFE coatings to be of prevailing extreme character, especially pronounced for HDPE coatings. The examination of these relationships led to conclusion that minimal J values were achieved at absolute values of σ_{eff} = ca.$\pm(3-5) \times 10^{-5}$ C/m^2 which coincides with the tribocharge values of σ_{eff} generated during sliding of initial coatings[2].

The results obtained prove our predictions made in terms of the friction energy balance, and allow us to distinguish a second aspect of the effect that the thermal

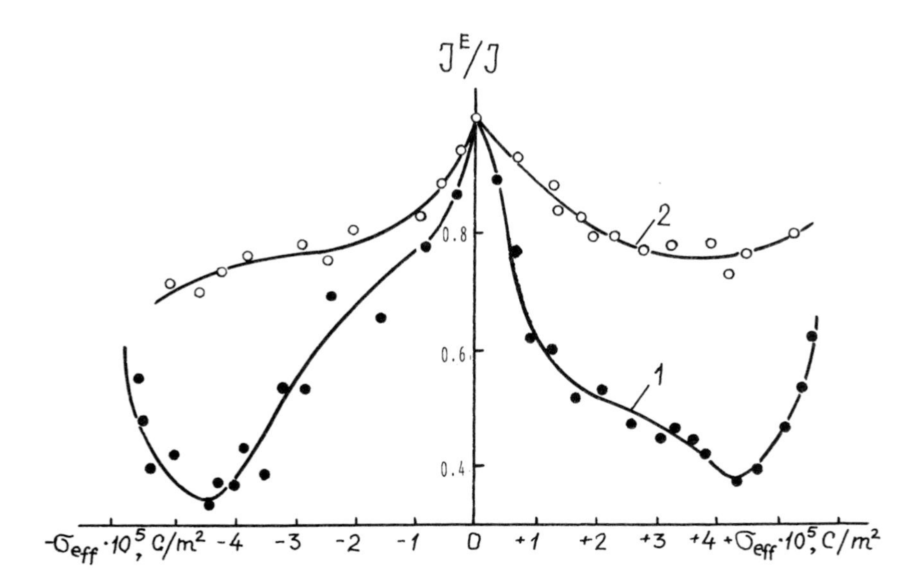

Figure 3 Relative wear rates during steady sliding vs. the sign and magnitude of ESCD for HDPE (1) and PCTFE (2) coatings treated in an electric field

electretization exerts upon the electrification processes, dispersion rate of polymer surface layers, and formation of friction transfer layers - the effect of the electret bulk charge field - Aspect II.

The differences in J values of the coatings and the mode of $J^E/J = f(E_e)$ dependences for the run-in period (the first hour of sliding) and steady sliding regime (Figure 2) can be explained by the simultaneous appearance of Aspect I - being dominant in the run-in period - and Aspect II - exerting a weaker but longer effect on the wear resistance of polymer coatings.

The duration of the modifying action of the volume charge field depends on the time of the electret charge relaxation; this time being dependent on the polymer material's dielectric properties and electric conductivity, and the conditions of sliding; it may be from several hours to several years[6].

Effect of Electron Beam Treatment

Variation in the properties of surface layers of polymer materials caused by bombardment with a non-penetrating electron beam are related to the radiation physicochemical transformations and formation of volume electret charge (radioelectret state).

The investigations conducted showed that electron treatment leads to considerable variations in the friction characteristics and physicomechanical properties (microhardness) of the coatings.

From data in Table 1 one can see that longer treatment periods t lead to reduced values of ESCD σ_{eff}. The wear rate J measured in the steady sliding regime reduces in an extreme mode and the friction coefficient f grows.

Microhardness H increased for cross-linkable polymers (HDPE and PCA) and decreased for degradable polymers (PCTFE and PP1). The results of X-ray analysis indicated the degree of crystallinity of HDPE and PCTFE coatings to decrease 5-11%, while for PCA and PP1 no significant variations were observed.

Much higher values of σ_{eff} could be achieved for non-polar and slightly polar polymers possessing high dielectric properties (HDPE and PCTFE) by the treatment under consideration, as compared with highly polar polymers (PCA and PP1).

The reduction in σ_{eff} values with increasing t is probably associated with a stronger effect of the injected charge scattered owing to the radiation-excited electric conductivity.

The experiments conducted showed that in most cases the highest value of ESCD for irradiated polymer coatings could be registered at t values within which range minimum values of J are the case. This is especially obvious with PCTFE coatings possessing high radiation stability.

It was of interest to understand the effect of electron energy of the treatment which determines the average length of run R in the polymer and, correspondingly, the thickness of the modified coating layer on variation of the coatings' properties.

Table 1 Effect of electron treatment duration t on polymer coating properties (E=75 keV)

Polymer	Value of t, ks	Charge density value σ_{eff}, μC/m^2	Values of $J \times 10^{10}$ & f	Microhardness value H, MPa
HDPE (p=0.5 MPa)	0	0	7.2 / 0.22	42.0
	0.1	28.5	0.5 / 0.43	43.0
	0.3	36.5	0.9 / 0.62	45.0
PCA (p=0.5 MPa)	0	0	15.4 / 0.39	132.6
	0.1	3.0	6.1 / 0.55	139.4
	0.3	1.1	7.3 / 0.68	143.0
PCTFE (p=0.25 MPa)	0	0	46.4 / 0.31	91.3
	0.1	39.1	30.9 / 0.31	98.2
	0.3	11.8	41.0 / 0.30	88.0
PPl (p=0.1 MPa)	0	0	65.6 / 0.32	160.1
	0.1	7.0	59.1 / 0.39	151.4
	0.3	2.2	52.2 / 0.43	142.3

Table 2 Effect of electron treatment energy on HDPE coating properties

Energy value E, keV	Value of time t, ks	Dose D, Mrad	Values of Δh, μm and of f	Microhardness H, MPa
0	0	0	5.8 / 0.22	42.0
25	0.18	398	4.8 / 0.26	43.8
	0.3	664	4.7 / 0.27	47.1
	0.6	1328	4.4 / 0.27	53.4
75	0.18	191	1.4 / 0.43	43.5
	0.3	318	1.3 / 0.62	45.0
	0.6	636	1.7 / 0.65	48.8

HDPE was used to establish that the reduction of E from 75 keV to 25 keV (R decreased from 75 μm to 12 μm) results in a marked decline in the effect of the electron treatment. The analysis of the data in Table 2 and Figure 4 revealed that reduction of E values at any t values causes a decrease in f, increases J values and worn layer thickness Δh of the treated coatings. The microhardness of the coatings was observed to rise, which correlates with the average dose rate D absorbed by layer thickness R. The latter was calculated from the known expression[6].

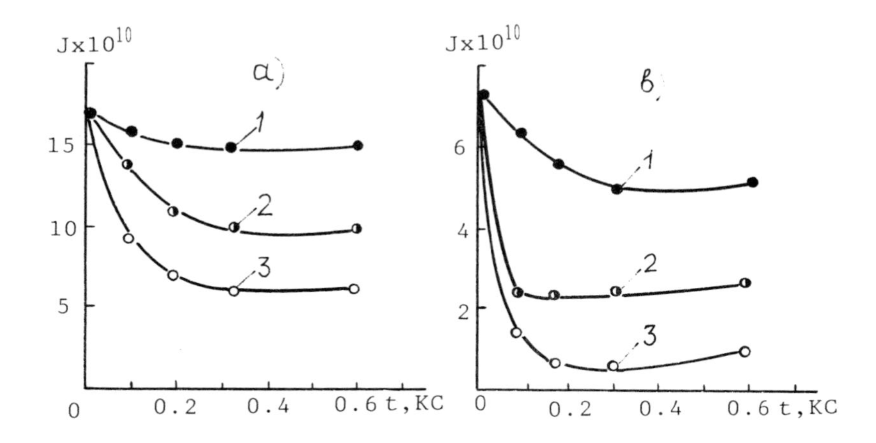

Figure 4 Effect of time during which electron beam acted upon coatings with energy 25 keV(1), 50 keV(2) and 75 keV(3) on wear rate in the first hour (a) and subsequent two hours (b) of sliding

The all-round consideration of the results obtained, taking into account the effects of radiation-chemical modification and the volume charge field, on the polymer properties suggested a conclusion that the variations observed in the coating properties are generally due to the chemical changes occurring in the molecular structure of the polymer surface layers.

However, the significant influence of the modified layer thickness R - where the volume charge is localized - on the wear rate and the friction coefficient at approximately equal absorbed doses D and equal extent of the structure transformations under the effect of the volume charge, could represent dependence of coating's properties on volume charge considerations.

The unusual manner of variations in the coating friction parameters J and f, depending on the electron energy, could be explained in terms of the wear theory by delimination process using the adhesion-fatigue model which describes the wear dynamics, and taking into consideration the processes of surface (adhesion) and subsurface (fatigue) fracture occurring simultaneously during friction[8,9].

4 CONCLUSION

It was established that electrophysical methods of modification, such as thermal treatment in an electric field, as well as treatment with a non-penetrating electron beam, can lead to essential alterations in the properties of polymer coatings. These procedures can be advantageously used for increasing the wear resistance of coatings.

REFERENCES

1. S.N. Postnikov, 'Electric Phenomena in Friction and Cutting,' Volgo-Viatskoe Book Publishers, Gorky, 1975.
2. A.F. Klimovich and V.S. Mironov, Soviet Journal of Friction and Wear, 1985, 6, 18.
3. Sh.M. Bilik, 'Metal-Plastics Friction Pairs in Machines and Mechanisms,' Mashinostroenie, Moscow, 1965.
4. A.R. Savkoor and T.V. Ruyeter, In 'Advances in Polymer Friction and Wear,' Plenum Press, New York-London, 1974, 5A, 333.
5. G. Fleischer, Schmierungstechnik, 1976, 7, 271.
6. G.M. Sessler (Ed.), 'Electrets,' Springer-Verlag, Berlin-Heidelberg-New York, 1980.
7. V.S. Mironov, A.F. Klimovich and A.P. Luchnikov, Soviet Journal of Friction and Wear, 1986, 7, 43.
8. V.A. Bely, A.I. Sviridenok, M.I. Petrokovets and V.G. Savkin, 'Friction and Wear of Polymer-Based Materials,' Nauka i tekhnika, Minsk, 1976.
9. N.P. Suh, Wear, 1977, 44(1), 1.

2.1.9
Investigation of Anti-friction and Protective Properties of Coatings Electrodeposited on Elastomers

A.S. Kuzharov, G.A. Danyushina, and I.E. Uflyand[1]

INSTITUTE OF AGRICULTURAL ENGINEERING INDUSTRY, ROSTOV-ON-DON 344010, RUSSIA

[1] STATE PEDAGOGICAL INSTITUTE, ROSTOV-ON-DON 344082, RUSSIA

1 INTRODUCTION

Rubber mixture volume modification and modification of ready-made technical rubber products' (TRP) surfaces are of the greatest practical interest for improving tribotechnical properties of TRP.

Surface modification has a number of advantages over volume modification since it gives the possibility to flexibly control TRP properties without interfering with the rubber mixture composition. Polymer and metal-polymer coatings play a prominent rôle among surface modification methods[1,2]. Such coatings, along with improving TRP friction characteristics, enhance their resistance to corrosive media and retain physical and mechanical properties of modified elastomers. Polymers and especially fluoroplastics having unique tribotechnical properties and a high thermal and chemical stability are used as components of antifriction and protective polymer coatings for TRP.

This paper is devoted to the technological development and investigation of tribotechnical and protective properties of thin layer composited fluoroplastic coatings for the most frequently used TRP produced from elastomers based on fluorine and nitrile rubbers.

2 RESULTS AND DISCUSSION

Modification of TRP Surface

The electrodeposition method[3], producing coatings of

uniform thickness on products with complicated configurations was used to deposit polymer coatings on TRP.

Since electrodeposited coatings are only possible on electrically conductive materials, it was obvious that there was only one way to impart electrical conduction to the same TRP - deposition of thin electrically conductive layers on the TRP surface.

Taking into consideration TRP chemical peculiarities, the technological process giving a high quality coating consists of the following operations: degreasing, etching, activation, chemical application of metal coating, electrodeposited polymer coating and thermal treatment of the coating (Figure 1).

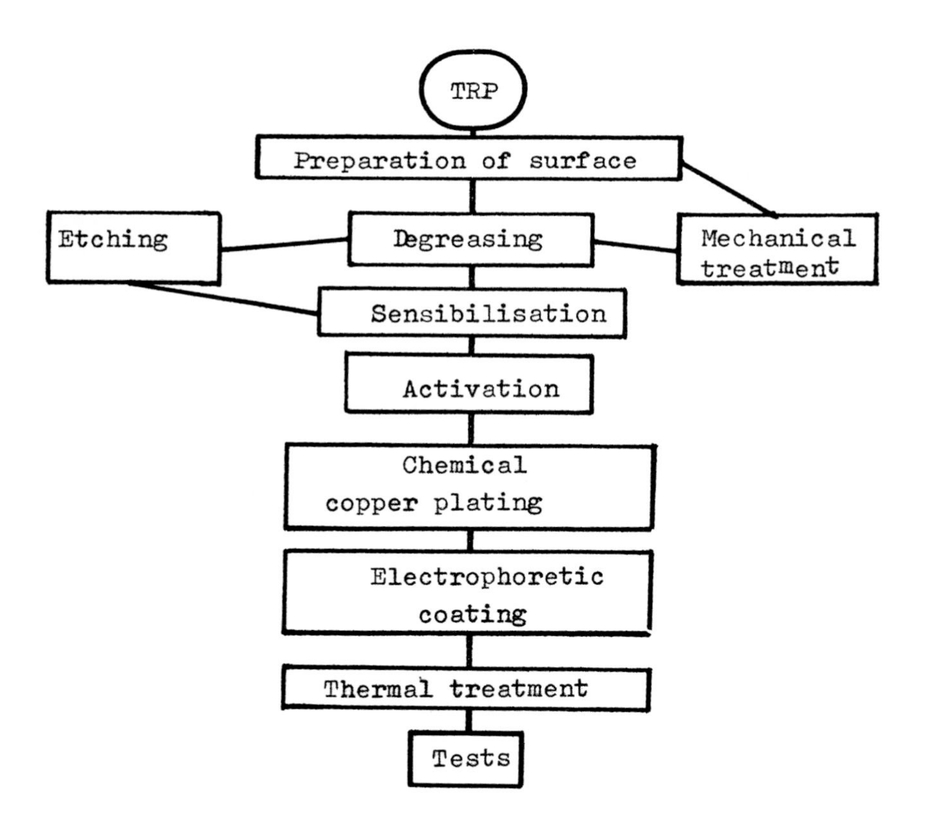

Figure 1 Schematic illustrating the technological process of TRP modification

The influences of each processing parameter - in particular surface preparation, electrodeposition and thermal treatment - over the efficacy of the coatings systems were thoroughly investigated.

Surface preparation, including degreasing and etching, predetermines the efficiency of the chemical application of the metal coating as well as the bond strength of the polymer coating with the base.

As a result of investigating changes in wetting angle and a post-treatment surface profile with different compositions, it was established that rubbers based on fluorine are best degreased in ethanol and those based on nitrile rubbers are best degreased in a mixture of butyl alcohol and acetone. Fluorine rubbers are best etched by a mixture of sulphuric acid and potassium bichromate and nitrile rubbers by a solution of ammonium chloride.

Particular attention in characterizing the electrodeposition process was paid to investigating compositional variations for electrodeposition since the exploitable properties of modified TRP are to a great extent determined by the composition of the obtained coating.

Compositions for electrodeposited coatings include film-forming agent and fluoroplastic materials.

Investigation of the chemical characteristics of the film-forming agent showed that epoxy resin and alkyl resin-based materials are ineffective to use in compositions to modify TRP; such coatings have high friction coefficients and low bending capacity (Table 1) which results in cracking of coatings even under minor deformations. At the same time coatings produced with a film-forming agent based on a mixture of oils have the necessary properties which give the possibility of using them as a component of coatings for TRP modification.

Strength properties of coatings are more sensitively dependent upon the chemical nature of the film-forming agent than on that of fluoroplastic. However the chemical characterization of the fluoroplastic does determine to a great extent the antifriction and antiwear properties of the modified TRP. Thus TRP with polyvinylidenefluoride- and polytrifluorochloroethylene-based coatings have low antifriction characteristics, i.e. friction coefficient is 0.15-0.20 whilst the maximum workload is not more than 10-15 MPa. Use of polytetrafluoroethylene and its copolymers gives acceptable tribotechnical and strength characteristics.

It is established that fluoroplastic mixtures, such as 4D and 4MD, in coating compositions are most effective. In this case an easily melted component of the fluoro-plastic mixture takes part in film-forming and a more temperature-resistant component, the antifriction filler, determines higher tribotechnical and strength properties of fluoroplastic-based coatings. The friction coefficient of such composited coatings is 0.02-0.07, workloads reach 30 MPa and bending capacity - not more than 5 mm.

Further improvement of elasticity as well as antifriction and protective properties of coatings can be obtained by adding a composition of 10% of different latexes, for example, latex 260 improves bending capacity by 1 mm.

Antifriction and Protective Properties of Modified TRP

Comparison of tribotechnical properties (friction coefficient dependence on: specific load, relative sliding speed and lubrication medium characteristics) of modified TRP with those of non-modified systems showed that modified TRP can work at specific pressures up to 20 MPa, relative sliding speed of 0.1 m/sec, producing a dry friction coefficient of ca. 0.1 and with lubricant about 0.04-0.06. Wear decreases in the latter case by 6-8 times.

Antifriction composited coatings deposited on TRP serve not only as a barrier preventing access of corrosive medium to the surface being protected but also regulating tribotechnical properties. In this connection protective properties of such coatings were investigated in oils (Table 2), sea and river water (Figure 2).

Table 1 Composition and Properties of Coatings

Sample	Components	Content, mass %	Deposition regime U,V	Deposition regime time, min.	Hardness regime T, ^{0}C	Hardness regime time min.	Bonding capacity mm
1	Primer FL-093 Fluoroplastic 4D Water	15 20 65	200	2	200	40	10
2	Enamel V-EP-2100 Fluoroplastic 4D Water	10 20 70	100	1	180	40	20
3	Enamel UP-1154 Fluoroplastic 4D Water	10 20 70	100	1	180	40	20
4	Primer FL-093 Fluoroplastic 4D Fluoroplast 32A (powd.) Water	10 10 10 70	200	2	180	40	10
5	Primer FL-093 Fluoroplastic 4D Fluoroplastic 4MD Water	10 15 10 65	200	2	200	40	5
6	Primer FL-093 Fluoroplastic 4D Fluoroplast 32A (powd.) Latex 260 Water	10 15 10 10 55	100	2	180	40	3

Table 2 Properties of modified TRP

Sample	Friction coefficient under pressure, MPa					Swelling in oil, %		
	5	10	15	20	25	24 h	480 h	960 h
Initial rubber						0.07	0.32	0.50
1	0.06	0.05	0.04	0.04	0.03	0.05	0.24	0.32
2	0.09	0.09	0.07	0.06	-	0.09	0.35	-
3	0.08	0.07	0.07	0.05	-	-	0.12	0.21
4	0.07	0.05	0.04	0.04	0.03	-	0.10	0.18
5	0.06	0.04	0.03	0.03	0.02	0.05	0.26	0.30
6	0.04	0.04	0.03	0.02	0.03	-	0.19	0.34

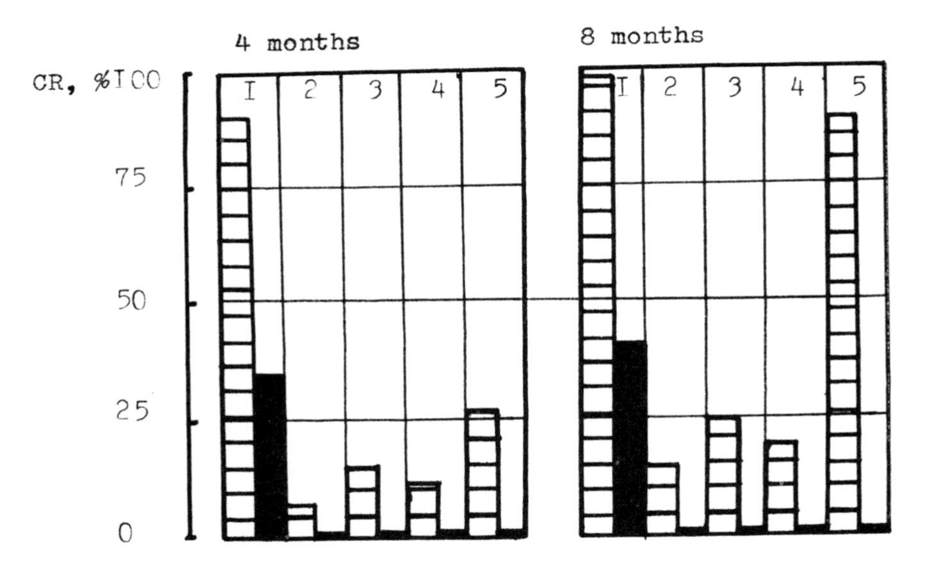

Figure 2 Corrosive resistance (CR) of steel (1), copper (2), aluminium (3), brass (4) and metallised elastomer IRP-1136 (5) with coating (blocked columns, seawater; ladder columns, river water)

3 CONCLUSIONS

The results obtained show that coatings based on a mixture of oils have a higher elasticity and better protective properties.

Comparison of results presented in this paper with the data for well-known plasmochemically modified TRP[4] and comparative tests (Table 3) show that electrodeposition of fluoroplastic materials on metallised TRP can be an effective means of controlling their tribotechnical and protective properties.

Table 3 Antifriction properties of modified TRP

Rubber	Recip-rocating motion, V=0.6 m/s	Trans-lational motion, V=0.1 m/s	Rotary motion, V=0.1 m/s, = 10 min.	
	Pressure, MPa			
	5	7.4	5	20
IRP-1136	0.8 0.6	0.8 0.54	0.7 0.4	destruct.
Modified rubber	0.12 0.1	0.14 0.1	0.12 0.08	0.9 0.06
Rubber modified by a plasmo-chemical method	0.12 0.08	- 0.32	0.25 0.2	0.2 0.15

REFERENCES

1. A.S. Kuzharov, 'Coordination Tribochemistry of Selective Transfer,' Dr.Sc. thesis, Rostov-on-Don, 1991.
2. G.A. Danyushina, 'Tribotechnical Properties of Antifriction Polymer Coatings Electrodeposited on Elastomers,'M.Sc. thesis, Rostov-on-Don, 1992.
3. I.A. Krylova, I.D. Kogan and V.N. Ratnikov, 'Electro-deposited Painting,' Khimiya, Moscow, 1982.
4. E.A. Duchovskoy, A.V. Semenov and A.V. Chomyakov, 'Modern Methods and Means of External Friction Measurements,' Proceedings of the All-Union Inst. of Phys.-Techn. and Radiotechn. Measurements, Moscow, 1977, pp. 60-64.

2.1.10
Fatigue of Gears: Surface Aspects

T.P. Wilks, G.P. Cavallaro, K.N. Strafford, P. French,[1] and J.E. Allison[1]

METALLURGY DEPARTMENT, UNIVERSITY OF SOUTH AUSTRALIA, ADELAIDE, SOUTH AUSTRALIA

[1] BIRRANA ENGINEERING, ADELAIDE, SOUTH AUSTRALIA

1 INTRODUCTION

Gears for power transmission and high load bearing applications are manufactured to strict quality assurance procedures in order to assure their reliability. This requires meticulous attention to materials, including steel cleanliness via steel making practice, design to optimize geometry, size and strength requirements (via material and process selection), process selection and control in forming and heat treatment operations, and surface engineering considerations such as surface hardening, surface finish and condition, and residual stresses.

Carburized steels are widely used for highly stressed machine parts such as gears, which are cyclically stressed and therefore resistance to fatigue fracture is a critical design factor.

An ongoing research programme by Birrana Engineering and the University of South Australia involves the evaluation of materials, process route and process control used in the manufacture of a range of gears and relates these factors to fatigue resistance, with the aim to reduce costs and improve quality and reliability.

The results of single tooth bending fatigue tests on production gears manufactured from EN39B steel with a carburised case showed that fatigue cracks were initiated at subsurface manganese sulphide inclusions within the carburized case, approximately 150-300 μm below the surface. Fatigue life was influenced by the size and position of initiating inclusions, maximum applied stress, and residual stresses resulting from heat treatment, carburising and glass bead peening operations.

This research has now been extended to include a carburised EN39B steel without glass bead peening, carburised AISI 8620 steel, and a steel X in the carburised and glass bead peened condition (details of grade and composition are restricted for commercial reasons).

EN39B and steel X initiated fatigue cracks from substrate MnS inclusions, whereas cracks in AISI 8620 were surface initiated and not associated with inclusions. The scatter of fatigue data was more evident in inclusion-initiated fatigue failures.

Extensive fractographic studies showed that the inclusion-initiated fatigue cracks initially grew in a transgranular mode, reaching a critical size before rapid fracture occurred through the carburised case in a brittle manner resulting in mixed intergranular/transgranular cracking. Subsequent failure of the core materials occurred in a ductile mode. The relatively rapid fracture of the carburized case, caused by the presence of a small fatigue crack greatly influences the fatigue resistance of the gears.

In this paper our own results are compared with published literature on fatigue crack initiation and growth in case carburised steels. The variables which influence crack initiation in, and toughness of the carburised case, such as carbon content, reheat treatments, embrittlement due to grain boundary segregation, grain size, retained austenite, residual stresses and tempering temperature are discussed in relation to our own findings.

2 EXPERIMENTAL

The experimental procedures are detailed in a previous paper[1]. Steel composition for the EN39B and AISI 8620 steels is given in Table 1. Single tooth bending fatigue tests were carried out on production gears and the load

Table 1 Steel composition

Steel		C	Si	Mn	Ni	Cr	Mo	S	P
En 39B	Min. Max.	0.12 0.18	0.10 0.35	- 0.50	3.80 4.50	1.00 1.40	0.15 0.35	- 0.05	- 0.05
AISI 8620	Min. Max.	0.18 0.23	0.15 0.30	0.70 0.90	0.40 0.70	0.40 0.60	0.15 0.25	- 0.04	- 0.035

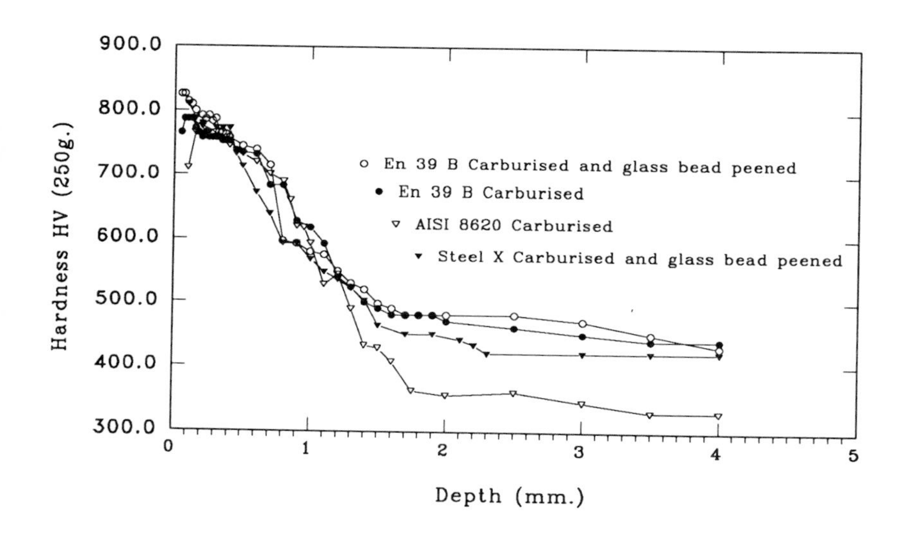

Figure 1 Hardness profile at root of gear tooth

applied took the form of a sine wave, cycling between zero and maximum load, at a frequency of 40Hz.

3 HARDNESS RESULTS

The hardness profile for the various gear material/ conditions tested is shown in Figure 1. Although the hardness of AISI 8620 material in the as-received condition (not carburised) is lower than that for EN39B and steel X, subsequent carburising resulted in a similar hardness profile in the case. The two carburised steels which were not glass bead peened (EN39B and AISI 8620) showed a reduced hardness near the surface (up to 200 μm depth) which is considered to be the result of decarburisation of the near-surface material during the reheat treatment. This effect was not apparent for the two steels which were glass bead peened (EN39B, steel X).

4 FRACTURE SURFACE EXAMINATION

Examination of the fracture surfaces of failed gear teeth, using optical and scanning electron microscopes (SEM), revealed the mechanisms of crack initiation and propagation in the various gear materials.

EN39B Carburized and Glass Bead Peened

As detailed in the Introduction, fatigue cracks were initiated at subsurface, elongated manganese sulphide (MnS) inclusions positioned 150-300 μm below the surface of the gear tooth root in the area of high stress concentration. Fatigue cracks initiated by the MnS inclusions grew in transgranular mode typically reaching depths of 400-600 μm below the surface of the gear, before the fracture appearance changed to a mixed intergranular/ transgranular mode, and propagated through the carburised case which was typically 1 to 1.4 mm below the gear surface. Beyond the carburised case, the fracture mode was ductile as the crack propagated by fast fracture.

Steel X Carburised and Glass Bead Peened

Fatigue cracks were initiated at subsurface elongated MnS inclusions, Figure 2, positioned 50-130 μm below the surface of the gear. Fatigue cracks grew in a transgranular mode, Figure 3, followed by a predominantly

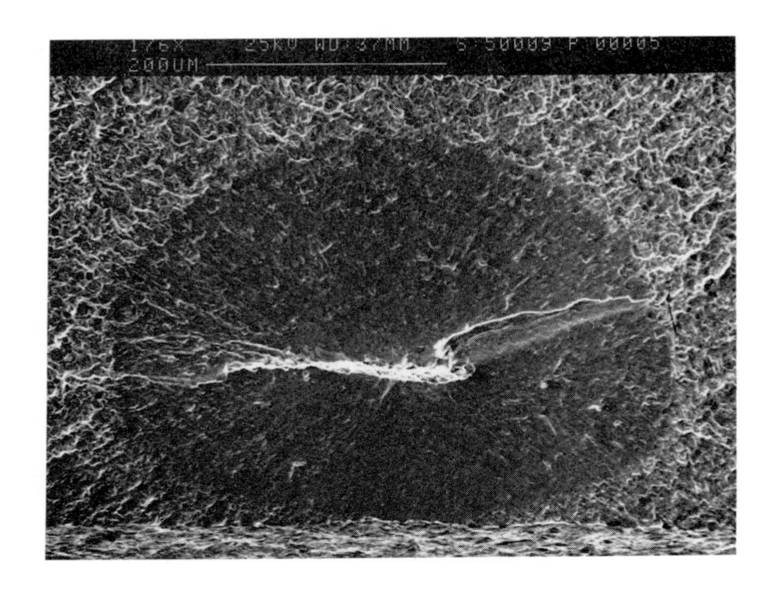

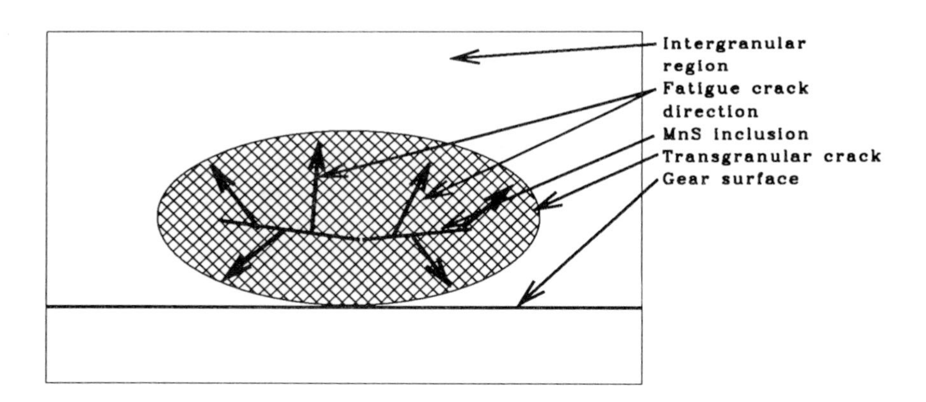

Figure 2 Fatigue crack initiation at MnS inclusion in steel X, (carburised and glass bead peened)

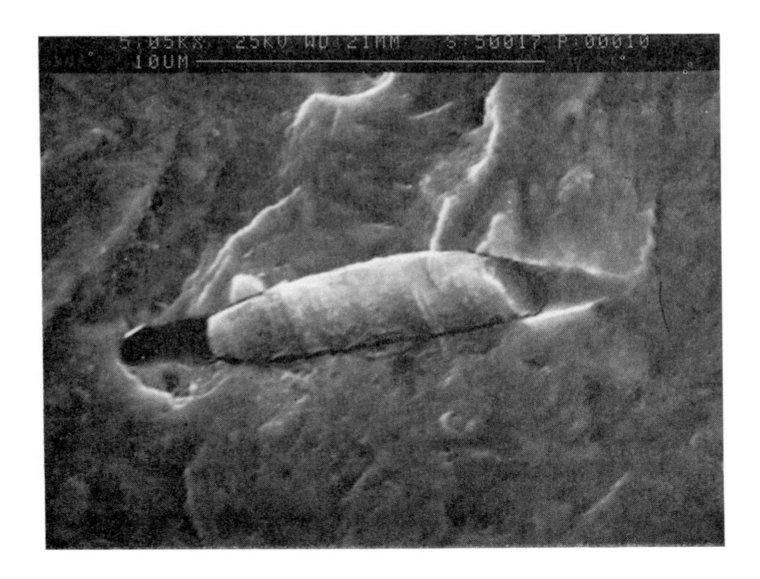

Figure 3 Transgranular fatigue crack, steel X, carburised and glass bead peened

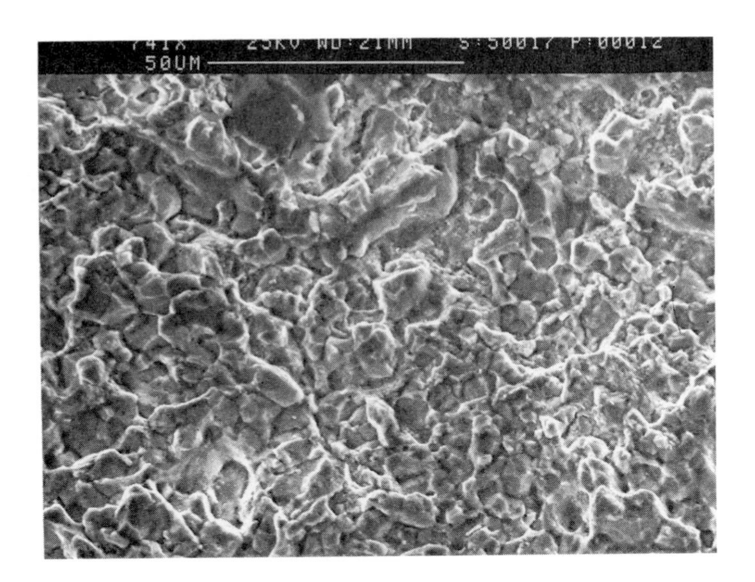

Figure 4 Intergranular fracture of carburised case, steel X

intergranular fracture which propagated throughout the carburised case, Figure 4. Transgranular cracking of a brittle nature was observed at the boundary of the carburised case and core material before fast fracture occurred in a ductile mode, Figure 5.

AISI 8620 Carburised

Fatigue cracks were initiated from the surface at the root of the gear tooth in the area of high stress concentration, Figure 6. Although the material contained elongated MnS inclusions they were not associated with fatigue crack initiation. Specific fatigue crack initiation sites were difficult to identify and crack propagation throughout the carburised case occurred in a predominantly intergranular mode. Beyond the carburised case, fast fracture occurred in a transgranular mode with evidence of crack branching at the fracture surface.

5 FATIGUE RESULTS

Fatigue results presented as maximum stress versus cycles to failure (S-N), are shown in Figure 7.

Steel X carburised and glass bead peened has the highest fatigue limit, above 1100MPa, with EN39B falling below 900 MPa with and without glass bead peening. The scatter of fatigue data is more evident in gear teeth which initiated fatigue cracks at inclusions, EN39B and steel X, with the AISI 8620 showing a well defined fatigue limit consistent with fatigue cracks initiated at surface irregularities.

The size, shape and position of inclusions which initiate fatigue cracks in martensitic steels have been shown to significantly influence fatigue life[2]. The fatigue lives of EN39B and steel X could be attributed to the size of the initiating MnS inclusions as well as their depth below the surface.

6 DISCUSSION

Appleman and Krauss[3] studied microcracking and fatigue in a carburised AISI 8620 steel subjected to three post-carburising heat treatments. Surface hardness dropped off due to decarburisation during the reheat cycle, and they

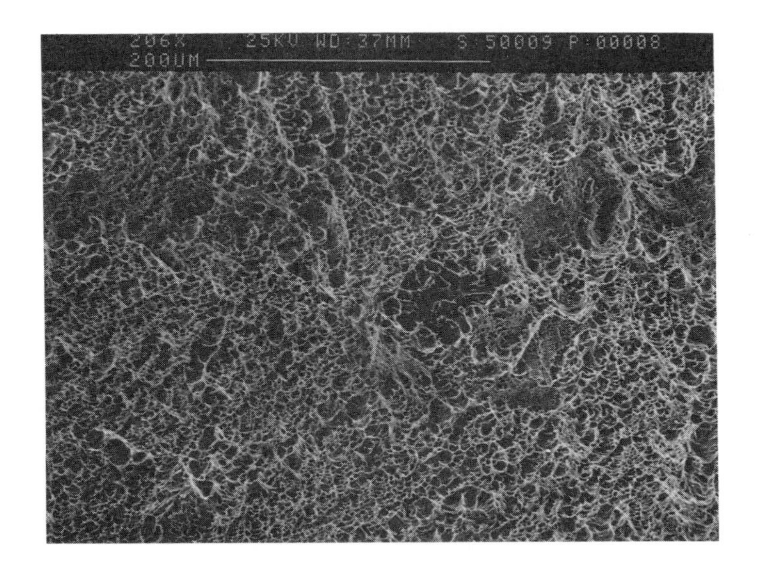

Figure 5 Ductile failure of core material, steel X

Figure 6 Surface fatigue crack in AISI 8620

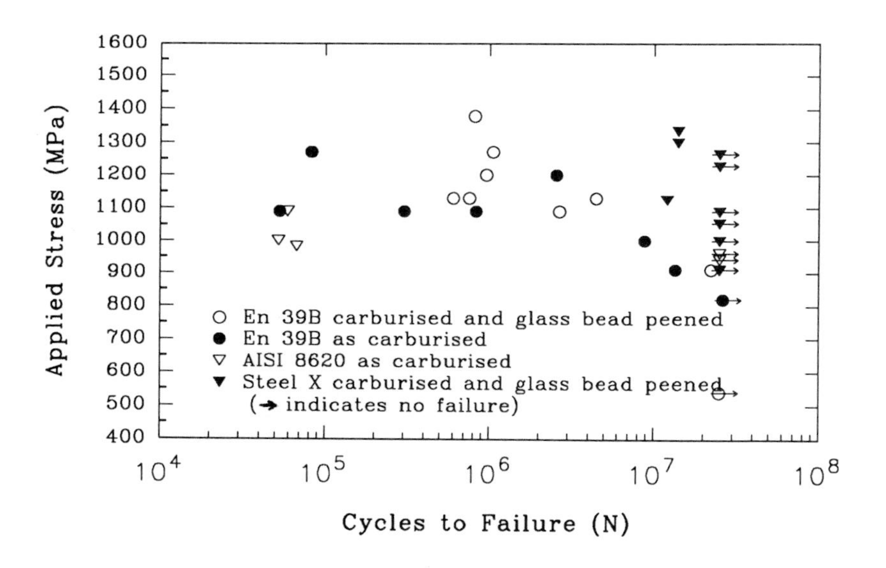

Figure 7 Maximum local stresses versus cycles to failure

concluded that the carburising atmosphere used in reheating was insufficient to maintain the surface carbon content established during carburising. Double reheat treatment after carburising resulted in superior fatigue properties than a single reheat treatment, which in turn was superior to a direct quench. The fracture surfaces in the case regions of the direct quench and single reheat specimens exhibited mixed intergranular/transgranular modes of fracture, similar to those observed in our own tests on AISI 8620. Appleman and Krauss[3] observed several specimens which showed a small featureless area of slow fatigue crack growth near the specimen surface. Interruption of the fatigue tests showed that once the crack reached a critical size, it propagated through the high carbon portion of the case very rapidly (within a few cycles) and was arrested by the relatively ductile core. Final failure then occurred after several hundred cycles. Although this area of slow fatigue crack growth (transgranular) was not apparent in the AISI 8620 tested in our own programme, tests on EN39B and steel X did result in such cracking from subsurface MnS inclusions, and a well defined boundary into the mixed intergranular/ transgranular mode. Multiple fatigue crack initiation observed in our own tests on AISI 8620 is consistent with initiation from surface irregularities.

Krauss[4] carried out Auger electron analysis to show that carbon and phosphorus were present on the intergranular fracture surfaces of the case at concentrations above the bulk material. Marcus and Palmberg[5,6] among the first to apply Auger analysis to embrittled steels, showed that Sb segregates to austenite grain boundaries of an embrittled AISI 3340 steel[5] and that Sb and P were concentrated within one or two atomic layers of the austenitic grain boundaries. They concluded[7] that P was a more effective embrittling element than Sn or Sb and that the presence of Ni and Cr together in a low alloy steel produced more Sb, Sn or P segregation than does the presence of either Cr or Ni alone. Krauss[8] observed that phosphorus concentrates in very thin layers at austenite grain boundaries of the order of atomic dimensions, and could only be detected by Auger electron spectroscopy (AES). Analysis showed that phosphorus was present at austenitic grain boundaries in as-quenched steels[9,10], thus tempering was not required to cause segregation.

Pacheco and Krauss[11] fatigue tested SAE 8719 steel (0.19% C; 1.06% Mn; 0.52% Cr, 0.5% Ni; 0.17% Mo) plasma and gas carburised with direct quenching and reheat treatments. The fatigue fractures typically showed four zones; an initiation zone, a transgranular region of stable fatigue crack growth, an unstable fracture zone through the case (mostly intergranular) and a ductile overload zone through the core. Fatigue crack initiation in the gas carburised specimen was transgranular and associated with surface discontinuities. Once cracks were initiated, they developed small plastic zones which caused stable transgranular crack propagation until the low fracture toughness of the high carbon case was exceeded.

Although austenitic grain size has been used as the microstructural parameter which correlates with high-cycle fatigue performance, it is the transformed martensite/ austenite microstructure which controls fatigue crack initiation and propagation. Finer parent austenite transforms to finer mixtures of tempered martensite and retained austenite with increased resistance to low strain plastic deformation due to higher elastic limits. Retained austenite in the case microstructures should also be minimised and finely dispersed for good high-cycle fatigue performance[11].

Magnusson and Ericsson[12] studied the fatigue fracture surfaces of carburised specimens to establish whether the intergranular cracking of the case was in fact a true fatigue crack or not. The microstructures of the case hardening steel (637 M17), directly quenched, were martensite in the case and a bainite-pearlite mixture with a small amount of ferrite in the core. Fatigue cracks were initiated at grain boundaries or slag inclusions and extended to depths of 100μm. Calculations[12] showed that the final fracture of the carburised layer occurred when $\Delta K \sim 12 MPa\sqrt{m}$.

Panhams and Fournelle[13] studied the fatigue resistance of a carburised steel as a function of surface retained austenite and surface residual stress. Refrigeration treatment of the carburised steel decreased the surface retained austenite, increased the surface residual stress and decreased the endurance limit for the steel. Three fractured regions were observed; fracture through the case (predominately intergranular), fatigue fracture through the core, and final fracture through the core. Szpunar and Bielanik[14] also looked at the effect of retained austenite on fatigue crack propagation in carburised cases and concluded that retained austenite can accelerate or hinder crack propagation, depending on the load applied. They found the stress field in front of the crack could lead to the transformation of retained austenite to martensite, thus consuming some of the work done by external forces, which diminishes crack propagation. Lee and Ho[15] found that retained austenite, as well as large grain size, had an adverse effect on the fracture toughness of the carburised case in an AISI 8620 steel. A tempering temperature of 500^{0}C provided maximum K_{Ic} values (115 MPa$\sqrt{m}$) however, lower tempering temperatures (150-300^{0}C) had little influence on fracture toughness which remained similar to the as-quenched value at about 105 MPa$\sqrt{m}$.

The AISI 8620 steel tested in our own programme was tempered at only 170^{0}C which would not significantly influence the fracture toughness of the as-quenched case. Tempering at a higher temperature would allow transformation of unstable retained austenite to tougher bainite and the precipitation or fine carbides to increase case hardness and overall toughness.

The formation of martensite structures in carburised steel is a function of hardenability[16] and alloy carburising steels are selected to ensure good case and core hardenability. However, because of alloying, quenching practice and section size, varying amounts of non-martensitic transformation products of austenite may form in the case and core. Most carburising steels are tempered at lower temperatures between 150-200^{0}C to preserve the high strength and hardness but reduce residual stresses and aid carbide precipitation, however, there is no change in austenite content from that retained in the as-quenched condition.

7 CONCLUSIONS

1. Steel X carburised and glass bead peened had the highest fatigue limit (above 1100 MPa) followed by AISI 8620 carburised (960 MPa), with EN39B carburised steel falling below 900 MPa with and without glass bead peening.

2. In EN39B and steel X fatigue cracks were initiated at subsurface (50-300μm depth) MnS inclusions within the carburised case. Fatigue crack growth was transgranular and reached a critical size (400-600μm depth) before changing to intergranular/transgranular fracture of the case. Fast fracture of the core material occurred in a ductile mode.

3. In AISI 8620 steel fatigue cracks initiated at the gear surface which propagated through the case in a predominantly intergranular mode, with core material cracking in a more ductile transgranular mode.

4. The scatter fatigue data was more evident in steels which initiated fatigue cracks from inclusions (EN39B, steel X).

8 FUTURE WORK

The relatively rapid fracture of the carburised case caused by the presence of a small fatigue crack, greatly influences the fatigue resistance of the gears. In this discussion we have considered the variables, as identified in published literature, which influence the toughness of the case material and therefore its ability to accommodate small fatigue cracks initiated on or near the gear surface. Although the inclusions responsible for fatigue crack initiation (EN39B, steel X) are inherent in the materials, thus limiting our influence over the initiating mechanism, there is scope to alter, via heat treatment, the properties of the matrix material which surrounds the inclusions. Similarly if the fatigue cracks in AISI 8620 are initiated from surface microcracks resulting from previous heat treatment cycles , effort should concentrate on avoiding their occurrence, or removing them.

The future research programme will concentrate on the following aspects:-

(A) Establish whether fatigue crack initiation in AISI 8620 is caused by the presence of surface microcracks formed during previous heat treatment.

(B) Estimate the levels of retained austenite in the case of the gear materials, and equate to fatigue performance.

(C) Consider the feasibility of raising the tempering temperature (presently 170^{0}C) in order to increase the fracture toughness of the case material, and establish the resulting fatigue performance of the gears.

(D) Perform Auger spectroscopic analysis of fracture surfaces to identify possible embrittling species at the grain boundaries in the carburised case (for example, phosphorus). If so, look at avoiding segregation via revised heat treatment procedures.

ACKNOWLEDGEMENTS

The authors would like to acknowledge Birrana Engineering Limited South Australia, for supporting this research programme. They would also like to express gratitude to Dr. Peter Slattery and Mr. Len Green, Department of Metallurgy, University of South Australia for valuable discussions and assistance.

REFERENCES

1. T.P. Wilks, G.P. Cavallaro and K.N. Strafford, "Fatigue of Surface Hardened Gears," Proc. of Int. Conf. on Surface Engineering: Practice and Prospects, March 1991, Adelaide, South Australia.
2. T.P. Wilks, PhD. Thesis, Newcastle University, U.K., 1985.
3. C.A. Appleman and G. Krauss, Met. Trans., 1973, 4, 1195.
4. G.Krauss, Met. Trans. A, 1978, 9A, 1527.
5. H.L. Marcus and P.W. Palmberg, Trans. TMS-AIME, 1969, 245, 1664.
6. P.W. Palmberg and W.L. Marcus, Trans. ASM., 1969, 62, 1016.
7. H.L. Marcus, L.H. Hackett (Jr) and P.W. Palmberg, ASTM STP, 1972, 499, 90.
8. G. Krauss, "Microstructure, Residual Stresses and Fatigue of Carburised Steels," Third Int. Sem. Quenching and Carburising, Melbourne, Australia, 1991.
9. O. Ohtani and C.J. McMahon (Jr), Acta Met, 1975, 23, 337.
10. T. Ando and G. Krauss, Met. Trans. A, 1981, 12A, 1283.

11. J.L. Pacheco and G. Krauss, "Microstructure and High Bending Fatigue Strength in Carburised Steel," Proceedings of the International Conference Carburising, Processing and Performance, 12-14 July 1989, Lakewood, Colorado, USA, pp. 169-190, ASM Publication, 1989.
12. L. Magnusson and T. Ericsson, "Initiation and Propagation of Fatigue Cracks in Carburised Steel," Proc. of Conf. Heat Treatment '79, The Metal Society, 1979.
13. M.A. Panhams and R.A. Fournelle, J. Heat Treating, 1981, 2, (1), 54.
14. E. Szpunar and J. Bielanik, "Influence of Retained Austenite on Propagation of Fatigue Cracks in Carburised Cases of Toothed Elements," Conf. Proceedings Heat Treatment '84, May 1984, London, The Metals Society.
15. S.C. Lee and W.Y. Ho, Met. Trans. A, 1989, 20A, 519.
16. U. Wyss, Harterei Technisehe Mitteilungen, 1988, 43, 27.

Section 2.2 Corrosion Control

2.2.1
Coatings for Improved Corrosion Resistance

K. Natesan

MATERIALS AND COMPONENTS TECHNOLOGY DIVISION, ARGONNE NATIONAL LABORATORY, ARGONNE, IL 60439, USA

1 INTRODUCTION

A number of advanced technologies that convert coal to clean fuels for use as a feedstock in chemical plants and for generation of electric power are being developed.[1] In general, coal conversion processes result in a complex, multicomponent, multiphase mixture, the composition of which depends on a number of factors that include feedstock chemistry and operating conditions. Further, conversion of coal releases a wide variety of contaminants from coal feedstock, char, ash, and sulfur/chlorine/alkali-containing species. The manner in which these contaminants are released, transported in the gas phase, and subsequently deposited on relatively cooler metallic surfaces can significantly affect the performance of several functional components.

The use of different coals under a variety of combustion conditions, ranging from widely varying combustion gas chemistry to widely varying nucleation, growth, and condensation of particulates, can have a direct effect on material degradation in these systems. Contaminants such as alkali, chlorine, and sulfur vaporize during gasification and combustion and eventually condense on metal surfaces, removing the protective layer from those surfaces. Because of diverse processing conditions (temperature, pressure, air/fuel ratio, etc.) involved in the conversion of coal to gas, a thorough understanding of the influence of various gaseous species and particulates on material degradation is needed to enable selection of adequate materials and coatings for a reliable system.

2 COAL-FIRED SYSTEMS

Several coal-fired systems are of interest for energy production. In conventional pulverized-coal-fired (PC) boilers, the high temperatures and the excess air needed for complete combustion result in the generation of significant amounts of SO_2, SO_3, and HCl, in addition to alkali-sulfate and alkali-chloride vapors. The condensation of these volatilized constituents and deposition of ash constituents lead to fouling and corrosion of evaporators and steam superheaters. In addition, the waterwalls of combustion chambers are subject to attack by a mixture of coal ash and alkali compounds. While the current superheater systems operate at steam temperatures in the range of 538-566°C and pressure of 2400 psi, efforts are underway in the United States and other countries to develop advanced steam cycles that operate at a steam temperature of 650°C and pressure of 5000 psi. Recently, the U.S. Department of Energy has proposed to develop a high-performance power system (HIPPS) in which a very-high-temperature furnace will be used for combustion of coal and the heat of combustion will be transferred to a working fluid (air) via heat exchangers. The high-temperature air, with auxiliary heat input, is expected to reach 1370°C at the inlet to a gas turbine.

Coal gasification technologies emphasize production of intermediate-Btu syngas for use as a feedstock for production of chemicals or is burned to provide industrial process heat, steam, and/or electricity. Because gasification occurs at a high temperature, 10-15% of the energy in the coal is converted to sensible heat in the raw gas, the recovery of which requires large heat exchangers. Apart from the size of the units, the material selected must withstand a hostile environment that contains low partial-pressure oxygen (pO_2) and moderate-to-high partial-pressure sulfur (pS_2), and coal slag condensates.

Combustion of coal in a fluidized bed is widely considered a viable process for producing electric power and generating industrial process steam. In power-generating applications, tubes carry a working fluid (either steam or air) that eventually drives a turbine. In steam cycles, the fireside surfaces of the superheater tubes reach 550-700°C, whereas in air cycles the tube temperatures can be as high as the bed temperature, i.e., 900°C. In a coal-fired open-cycle magnetohydrodynamic (MHD) system, the working fluid is utilized on a once-through basis and consists of multiphase coal-combustion products. In the present MHD concept, the combustor walls will be exposed to a reducing environment with coal slag deposits, and the channel electrodes and insulators will be exposed at high temperatures to an environment of ionizing gas, molten slag, and sulfated seed material. Downstream components will be exposed to deposits of K_2SO_4 and/or ash constituents and an oxidizing-gas environment.

Other systems/components of interest are direct coal-fired gas turbines and indirectly-fired turbines that use the effluents from coal gasifiers in combined-cycle plants and from FBC systems in cogeneration plants. Performance requirements for the materials in these turbines are much more demanding, life expectancy is longer, and repair/reuse/cost considerations are much more stringent than those for aviation turbines.

Conversion of coal into clean energy in any process,

either by conversion to gaseous and liquid fuels or by direct combustion, requires materials that can withstand high temperatures in aggressive environments. Construction materials often are the limiting factor in the design of vessels and components for containment of the processes. Analysis of several of these systems indicate that the gas environment in different systems can vary widely in terms of oxygen activity, contaminant chemistry, temperature range, types of deposits and particulates, and particulate velocity range. Several modes of material degradation that include oxidation, sulfidation, hot corrosion, deposit-induced corrosion, and erosion have been identified. Further, key variables and possible rate-limiting steps for the degradation of materials in coal-fed energy-producing systems can be identified (Table 1). Details of the mechanisms for metal wastage exposed to environments in several coal-fired systems are discussed elsewhere.[1] Furthermore, the classes of materials used for construction of different components in these systems are also wide ranging. For example, relatively inexpensive carbon steel and low-alloy steels are used for waterwalls, economizers, and evaporators; austenitic alloys are used for steam superheaters and reheaters; high-chromium/aluminum-containing iron-base alloys are used for sulfidizing conditions; and nickel- and cobalt-base alloys are used for service in hot corrosion environments. As a result, the development of coatings for use in these systems must focus on a given application and the decision to use a coating is based on several technical and economic factors, some of which are listed in Table 2.

3 DEVELOPMENT AND APPLICATION OF COATINGS

Several coating approaches are being developed to resist attack in coal-fired environments and thereby minimize corrosion of underlying substrate alloys and extend the time for onset of breakaway corrosion. In general, coating systems can be classified as either diffusion or overlay type, which are distinguished principally by the method of deposition and the structure of the resultant coating-substrate bond. The coating techniques examined are pack cementation, electrospark deposition, physical and chemical vapor deposition, plasma spray, and ion implantation. In addition, ceramic coatings are used in some applications.

Pack-Cementation Coatings

Pack cementation is a process in which the component to be coated is fully immersed in a powder mixture of a coating element, an activator, and an inert material. The pack with the component is heated to temperatures between 750 and 1100°C for time periods of 4-24 hr in a protective atmosphere. This coating process is fairly simple and inexpensive and large components can be coated with relative ease. However, the pack process lends itself to only simple coating compositions, generally of a single element, primarily because of thermodynamic limitations on transfer of different elements under the same temperature and activator

Table 1 Materials Degradation in Coal-Fired Systems

Phenomenon	Key Variables	Possible Rate-Limiting Step
Boiler-Tube Corrosion	Alkali content Chlorine level Temperature Fly ash	Alkali condensation Oxide-sulfate reaction
Substoichiometric Combustion	Oxygen, sulfur pressure Temperature Downtime condensate Alkali/slag deposit	Fracture of oxide scale Oxidation-sulfidation Pitting/crevice formation
FBC In-Bed Corrosion/Erosion	Bed chemistry Local particle velocity Particle loading and size	Oxidation/sulfidation Arrival rate of particles Fracture of surface scales
Low-temperature Hot Corrosion	Temperature Salt-film thickness Temperature gradient Sulfur/Alkali level	Sulfidation of transient oxides Transport of base alloy elements
Hot Corrosion/Erosion	Alkali level Temperature Particle size, loading, and velocity	Fracture of scale Sulfidation of Transient oxides Transport of base alloy elements

Table 2 Technical/economic Factors in Coating Application

Cost comparison with monolithic but expensive material
Cost comparison with cladding and/or coextruded materials
Chemical compatibility with environment
Chemical compatibility with substrate
Coating/substrate interactions over time
Ease of application/coating integrity/reproducibility
Joining of coated components
Thermal compatibility
Feasibility of repair/recoat in field
Replacement cost and downtime expense
Initial cost outlay

conditions. Attempts are being made to coat more than one element simultaneously, but this effort is still in early stages of investigation and the performance of these coatings in complex environments must be evaluated. Furthermore, during the coating process, the component to be coated is exposed to temperatures much higher than the service temperature and this can alter the microstructural and mechanical properties of the base material.

Substantial work has been conducted to evaluate the corrosion behavior of aluminized, chromized, simultaneously aluminized and chromized, sequentially aluminized/chromized, and siliconized coatings exposed to simulated oxygen-sulfur environments.[2-6] Coatings were applied on low-alloy (1.25 and 2.25 wt.%Cr) and 9 wt.%Cr ferritic steels and austenitic alloys, such as Type 304 stainless steel and Alloy 800. Baxter[3] has established that a minimum chromium concentration of 25 wt.% is required to achieve adequate corrosion resistance in sulfur-containing gasification atmospheres under thermal-cycling conditions. A minimum aluminum concentration in a range of 15-20 wt.% is required for adequate corrosion resistance of aluminized coatings.

Figure 1 shows cross sections of chromized 2.25 wt.%Cr-1Mo steel after exposure in a high-sulfur atmosphere under isothermal and thermal cycling conditions.[3] Although the original coating contained 23.1 wt.% chromium, the enhanced intergranular corrosion under thermal-cycling conditions was attributed to oxidation and sulfidation of chromium-rich carbides precipitated during the coating process. The attack on these carbides is especially severe when the component is cooled below the dew point of the gas, whereupon the liquid condensate containing sulfur and chlorine preferentially attacks the carbides and causes pitting and intergranular corrosion. Studies have been conducted on minimizing the degrading effects of chromium-rich carbides on coating integrity by eliminating chromium carbide precipitation and/or adding elements to the coating that reduce pitting.[4] Additions of vanadium to chromized coatings by co-diffusion of chromium and vanadium in the pack process eliminated precipitation of chromium-rich carbides at grain boundaries, thereby reducing pitting and intergranular corrosion of the material. It was concluded that coatings containing 25-40 wt.% chromium and 4-6 wt.% vanadium on a low-alloy-steel substrate had adequate corrosion resistance, based on short term tests. Long term testing of these coatings is required to establish the effects of exposure temperature, sulfur level in the gas, ash/slag deposit, etc. on coating performance.

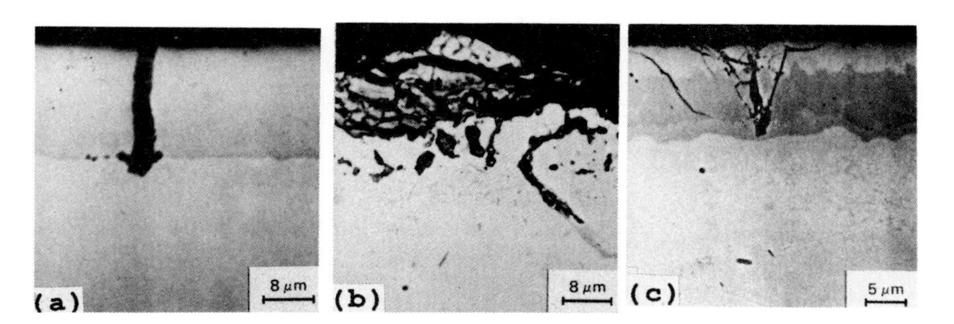

Figure 1 Cross sections of the chromized carbon steel after exposure in (a) a low-pS_2 gas under isothermal conditions, showing a breach in carbide-rich coating, (b) high-pS_2 gas under isothermal conditions, showing corrosive breakdown of carbide-rich coating, (c) high-pS_2 gas under thermal-cycling conditions, showing severe cracking of the coating.

Baxter[3] also concluded that pack-diffusion aluminizing and simultaneous aluminizing/chromizing processes produce coatings on low-alloy and carbon steels with good resistance to corrosion in simulated gasification atmospheres, but only under isothermal conditions. Sequentially chromizing/aluminizing produced coatings with physical defects, such as voids and cracks, and had poor corrosion resistance.

Coatings applied onto a T91 substrate by pack cementation showed that increased chromium and/or aluminum concentration is beneficial in resisting sulfidation attack, but that the integrity of the coating is strongly dictated by mechanical properties and by adhesion of the coating to the substrate rather than by chemical interactions with the exposure environment.[7] Figure 2 shows scale thickness and penetration data for the T91 alloy in uncoated condition and with different coatings after 500 h of exposure in a high-pS_2 gas environment at 650 and 500°C. For the uncoated alloy, corrosion losses (scale thickness plus penetration) at 500 and 650°C were ≈100 and 600 μm, respectively, which translate to ≈1.75 and 10.5 mm/yr, based on linear kinetics for the corrosion process. On the other hand, the chromized and chromized/aluminized coatings performed significantly better than the base alloy. Scale thicknesses were substantially lower (e.g., 40 μm, for a rate of 0.7 mm/yr at 650°C) at both exposure temperatures. The aluminized coating exhibited somewhat poorer performance in terms of void formation and cracking, which can be attributed to fabrication of the coating rather than to reactions occurring during exposure in a mixed-gas atmosphere.

The pack-cementation process has been used to examine codeposition of chromium/aluminum and chromium/silicon onto ferritic and austenitic alloys, and coatings with compositions of 25-35 wt.% chromium and 2-3 wt.% silicon have been reported.[5,6] Oxidation rates for these coated alloys have been lower than for those without coatings. Sulfur resistance of these coatings is not known at present.

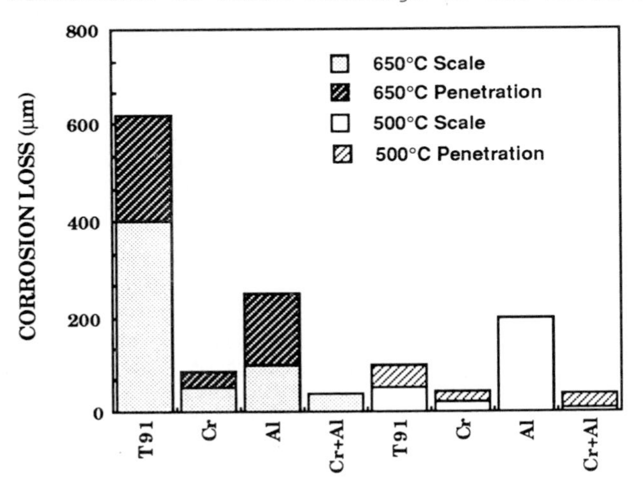

Figure 2
Corrosion loss data for T91 alloy in uncoated condition and with several coatings after 500 h of exposure to a high-sulfur gas environment at 500 and 650°C.

Chromized and aluminized coatings developed by the pack cementation process have also been examined for service in FBC environments.[8] In general, the results showed that the performance of aluminized coatings was poor. This is due more to difficulty in developing a crack-free coating than to exposure in an FBC environment. However, cracked regions, if present initially, exhibited accelerated oxidation and the coating integrity became poor.

Figure 3 shows scanning electron micrograph of cross section of chromized Alloy 800H after a 3000-h exposure. Also shown in the figure are the elemental concentration profiles across the coating to the substrate regions of the specimens. The coating interacted with the deposit, and a thin layer of corrosion reaction product was observed. Except for the interactions in the surface regions of the coatings, the elemental concentration profiles from X-ray analyses were similar to those for as-coated specimens.

Chromized coatings have also been tested for their resistance to coal-ash corrosion in superheaters and reheaters.[9] Results showed pitting attack of the coating at high concentrations of SO_2 and alkali sulfate deposit. Additional tests are needed to establish the mechanisms of pitting attack and to establish performance envelopes for application of these coatings in boiler fire side environments.

Aluminide coatings developed by the pack-cementation process have been used extensively as protection against hot corrosion of nickel- and cobalt-based superalloys in gas turbine applications. Corrosion of materials in the presence of liquid sodium sulfate, either by itself or in combination with sodium chloride, has been a problem in gas tur-

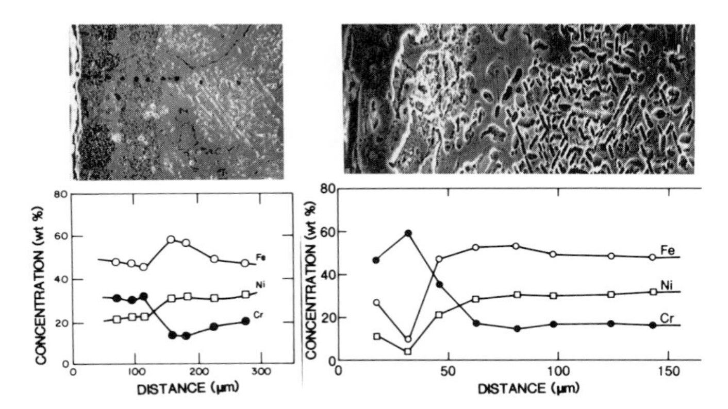

Figure 3 Scanning elctron micrographs and profiles of iron, chromium, and nickel concentration in chromized alloy 800H after a 3000-h exposure in the presence of $CaSO_4$ (left) and circulating-bed ash (right).

bines; this corrosion process has been termed "hot corrosion" to differentiate it from gas-phase sulfidation attack. Two types of hot corrosion have been identified: Type I, operative at 800-950°C, and Type II, operative at 600-750°C.

Type I hot corrosion can be divided into an initiation (or incubation) stage and a propagation stage. The process, in general, requires the presence of liquid sodium sulfate (melting point 884°C) on the metal surface. In the initiation stage, the protective oxide scale dissolves by a basic fluxing mechanism and the corrosion rates are generally low. In the propagation stage, with the protective oxide having been destroyed and not able to reform, the alloy is subjected to sulfidation by inward diffusion of sulfur, leading to accelerated corrosion rates.

Type II hot corrosion, also known as low-temperature hot corrosion, involves the eutectics of base-metal sulfates and sodium sulfate and, therefore, occurs predominantly at lower temperatures, especially in turbines that operate in the effluent of FBCs. Protective coatings examined for hot-corrosion resistance are simple aluminides, precious-metal aluminides, and overlay-type coatings. The former two are made by the pack process, whereas in the overlay process, the coating elements are applied directly onto the substrate by physical vapor deposition (PVD) or by the low-pressure plasma spray (LPPS) approach.

Figure 4 shows comparative data for corrosion of simple aluminides, aluminides with rhodium and platinum bond coats, and CoCrAlY overlay coating after exposure to a hot corrosion environment at 982°C.[10] The results show that the performance of a given coating is strongly influenced by the substrate alloy (more so in pack than in overlay coatings) and a bond coat of precious metal platinum (rather than rhodium) is beneficial in reducing corrosion in alloys such as IN 792 and IN 738. The overlay CoCrAlY coating on a

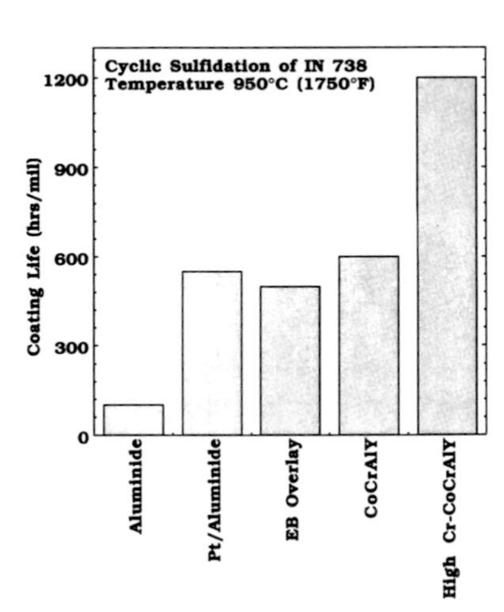

Figure 4
Coating life for several aluminide, precious-metal aluminide, and overlay coatings on nickel- and cobalt-base alloys exposed to hot corrosion environments.

cobalt-base MarM-509 alloy exhibited the best performance among all the coatings examined in this study. Extensive microstructural analyses of the tested coatings have been reported[17] and qualitative inferences have been drawn from the test results but a discussion of those results are beyond the scope of this paper. Corrosion test data obtained in the effluent of a pressurized FBC system showed that both nickel- and cobalt-base alloys were equally susceptible to accelerated corrosion. This susceptibility to accelerated corrosion was attributed to the presence of potassium in the FBC effluent.[16] Information on the performance of coating/substrate combinations under low-temperature hot- corrosion conditions is particularly lacking at present. Another study conducted to evaluate the performance of different coatings and coating methods on a given alloy substrate showed that the LPPS overlay of CoCrAlY, especially the high-chromium version, exhibited the longest life when compared with electron-beam overlay of CoCrAlY, precious metal aluminides, and simple aluminides (see Fig. 5).[11]

Electrospark Deposition (ESD) Coatings

The ESD process is a microwelding technique that uses short-duration, high-current electrical pulses to deposit an electrode material on a metallic substrate. A principal advantage of ESD is that the coatings are fused to a metal surface with a low heat input while the bulk substrate material remains at ambient temperature. This eliminates thermal distortions or changes in the metallurgical structure of the substrate.

Several coatings were applied on T22 and T91 substrate alloys by the ESD technique and tested in simulated gasification atmospheres that contained sulfur as H_2S.[7] Figure 6 shows a comparison of the corrosion-loss data obtained in

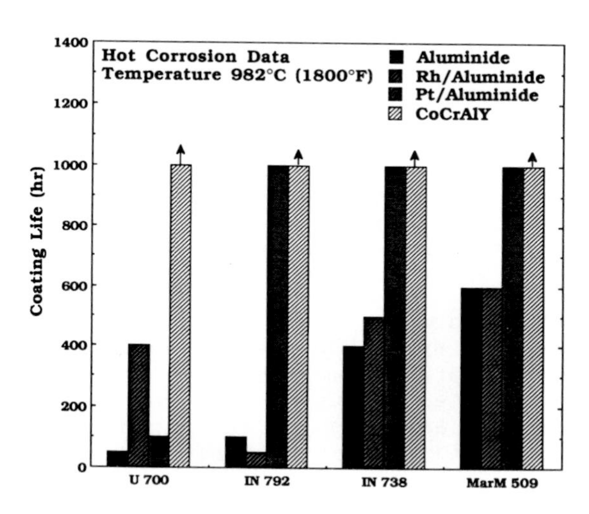

Figure 5
A comparison of coating life for several coatings on IN 738 exposed to hot-corrosion conditions.

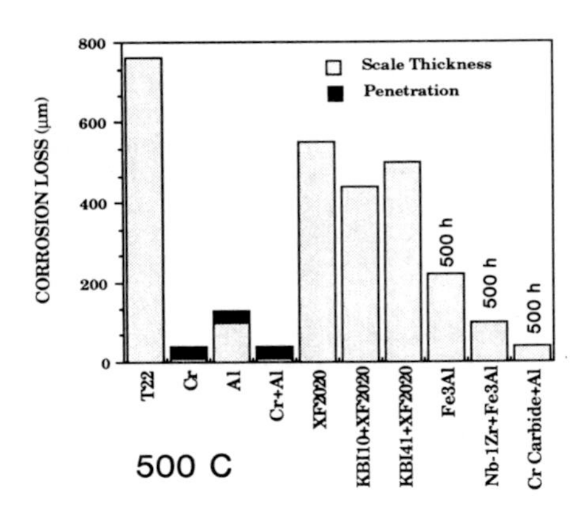

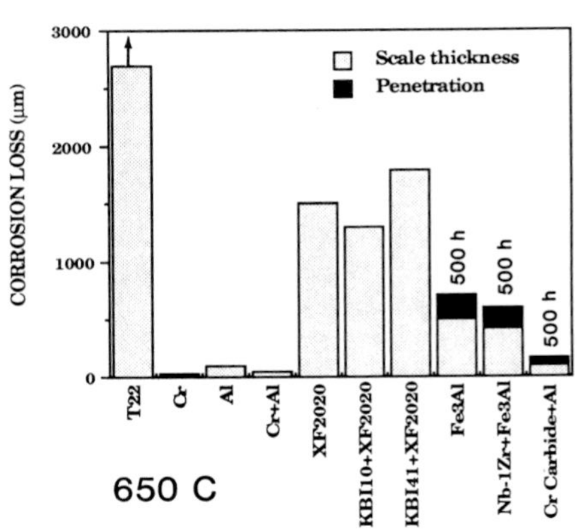

Figure 6 Corrosion loss data for T22 alloy in uncoated condition and with several coatings after 2000-h (unless marked otherwise) exposure to a high-sulfur gas environment at 500 and 650°C.

the uncoated condition and for several different coatings after exposure to a high-pS_2 environment at 500 and 650°C. The results showed substantial improvement for the pack cementation coatings and the ESD Cr carbide/Al coating in these reducing environments.

Plasma Spray Coatings

In a conventional plasma spray process, the coating material, in powder form, is melted in a plasma arc and propelled onto the target surface to form a coating. The method is amenable to development of coatings of complex compositions and large thicknesses but the porosity of the coatings is generally high. In addition, poor adherence of the coating to the substrate alloy can lead to spalling of the coating, especially under thermal- cycling conditions. Vacuum or low-pressure plasma spray, in which high particle and gas velocities are achieved, has been used to develop coatings with better corrosion resistance than those developed by conventional plasma spray. Interaction between the coating constituents and the underlying alloy elements is generally minimal and the coating, for all practical purposes, is similar to an overlay coating. However, the size limitation on the vacuum chamber makes coating of large components using this technique virtually impossible.

The resistance to sulfidation of several coatings (applied by low-pressure plasma spray) have been evaluated for application in coal gasification atmospheres.[12] Figure 7 shows cross sections of uncoated and Co-Cr-Al-Y coated IN 792 specimens after exposure in an oxygen/sulfur mixed-gas environment at 871°C. It is evident that the uncoated specimen formed liquid nickel sulfide and exhibited catastrophic

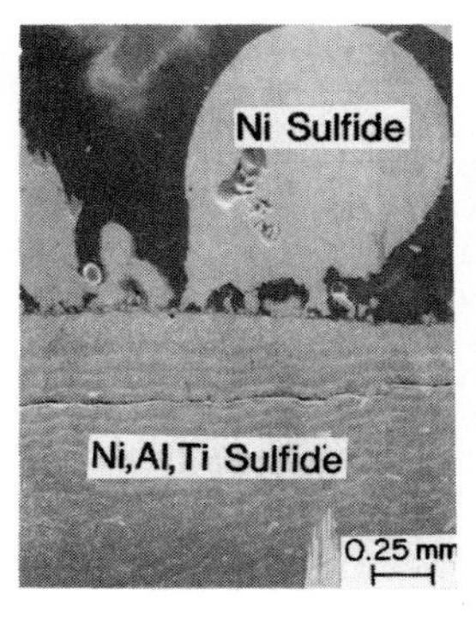

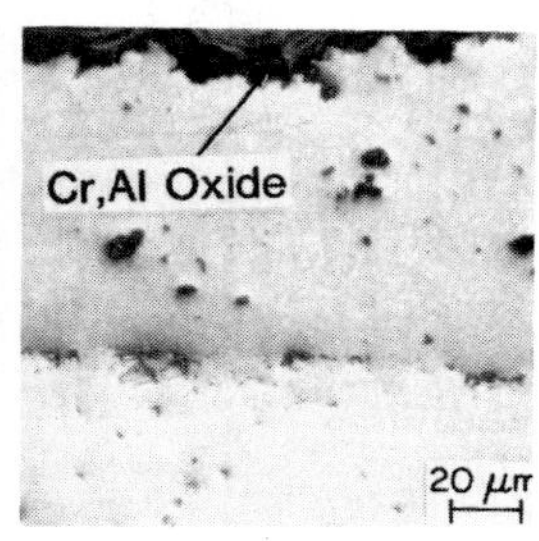

IN 792, uncoated **IN 792, Co–Cr–Al–Y coated**

Figure 7
The corrosion behavior of uncoated and coated IN 792 with Co-Cr-Al-Y after exposure in an oxygen/sulfur mixed gas at 871°C.

corrosion, whereas the coated specimen offered substantial resistance to corrosion by sulfidation.

Physical- and Chemical- Vapor Deposition

These involve vaporization of coating constituents from a source and subsequent deposition onto a substrate alloy. Several means for vaporization, such as, electron beam, plasma, and arc, have been used in the development of coatings. Even though PVD is a viable procedure for coating applications, lack of or poor adhesion of the coating to the substrate can be detrimental to coating performance. However, a number coatings developed by these techniques have been examined for their corrosion resistance and for protection of the underlying alloy for service in coal-fired systems.

Thin layers of silicon or silicon plus oxygen has been sputter deposited onto iron-base alloys, which were subsequently tested at 700°C in mixed-gas atmospheres that contained low pO_2 and high pS_2.[13] Thermogravimetric test data are shown in Fig. 8 for the uncoated and Si- or Si-plus-O-coated alloys. The uncoated alloy exhibited catastrophic sulfidation corrosion whereas the coated alloys exhibited virtually no sulfidation. Auger-electron- spectroscopy depth profiles for several of the tested specimens showed that enrichment of silicon and oxygen at the alloy surface acts as a barrier to migration of sulfur inward and cation transport outward of substrate, thereby minimizing sulfidation of substrate elements and reducing alloy corrosion rate. Even though substantial reduction in corrosion rates is possible by this approach, these techniques can, at best, be considered as research tools for optimizing coating compositions and morphologies and establishing viable coating/substrate combinations. The same coatings must be de-

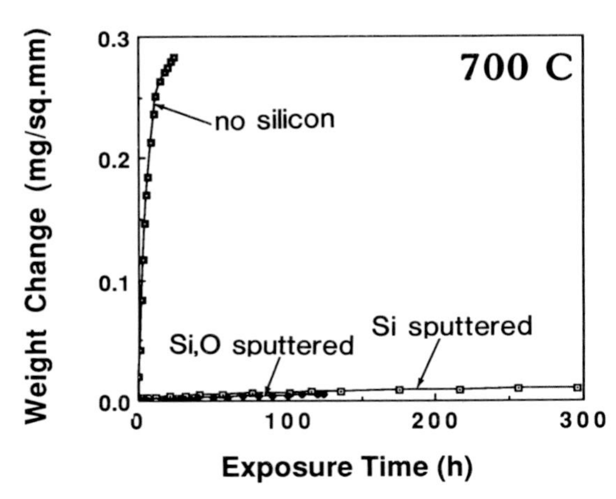

Figure 8
Thermogravimetric test data for Fe-25 wt.% Cr alloys with and without silicon or silicon-plus-oxygen-coating after exposure to a high-sulfur environment.

veloped by methods that are amenable to large components and at lower costs for wide-scale application in coal-fired systems.

Ion Implantation

Surface modification by ion implantation is predominantly a research tool and is used as such to evaluate the role of variations in coating composition, coating morphologies, adhesion characteristics, substrate/coating interactions, and coating response to gas phase and deposit layer constituents in the corrosion degradation of structural alloys and engineering ceramics.

Sulfidation resistance of iron-base alloys has been substantially improved by implantation of niobium on the surface.[14] In general, the growth rate of protective oxide scale is much higher in implanted alloys and is reflected in larger oxide grain size. Further, the presence of niobium and/or its oxide seems to act as a barrier to transport of sulfur inward and outward cation transport, thereby prolonging the time for onset of breakaway corrosion. Figure 9 shows typical scanning electron micrographs obtained from two different alloys, with and without Nb implantation, after exposure to a high-sulfur mixed-gas environment at 700°C. Reactive elements, such as Y, La, and Hf, have been implanted in structural alloys to modify the surface regions in chromia- and alumina-forming alloys to improve scale integrity and scale/metal adhesion, especially under thermal-cycling conditions. Most of the studies on surface-modified alloys have been conducted to improve the high-temperature oxidation resistance, and little data are available on the performance of these "coatings" in complex environments of interest in coal-fired systems.

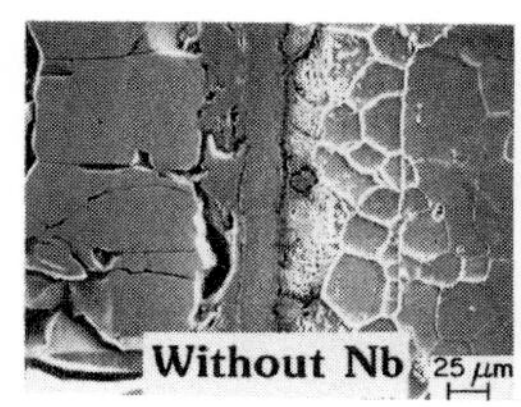

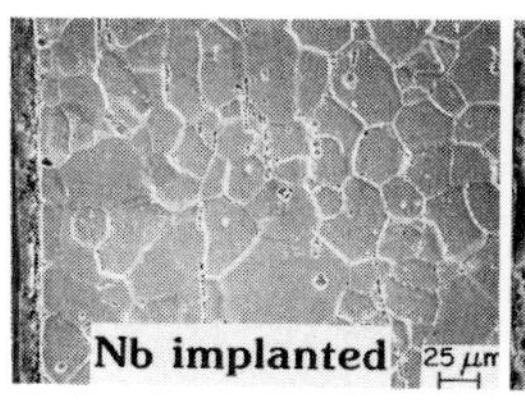

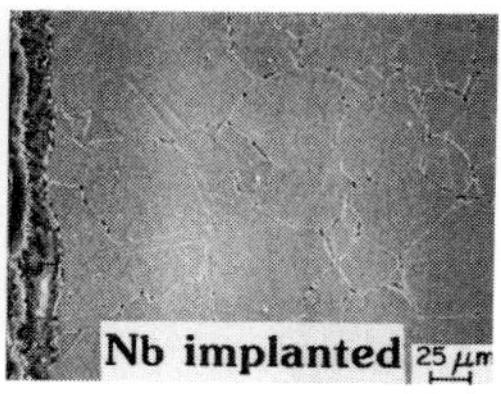

Fe-25Cr **Fe-25Cr-20Ni**

Figure 9
Scanning electron micrographs of cross sections of iron-base alloys with and without implanted niobium, after exposure to a high-sulfur environment at 700°C.

Ceramic Coatings

These coatings have been extensively discussed in the literature but the application of these coatings in coal-fired systems is minimal. One major application for the ceramic coatings, called "thermal-barrier coatings," is in gas turbines in which a coating with a low thermal conductivity is used to thermally shield the blade material from hot gases. Several reviews on different aspects of thermal-barrier coatings are published.[15-17] A bond coat of MCrAlY is applied onto the component by either the PVD, pack process, or plasma spray technique. Subsequently, a ceramic coating,usually Y_2O_3-stabilized ZrO_2, is applied onto the bond coat by the air plasma spray technique. Coating spallation is the major mode of failure of thermal barrier coatings and research to develop better adherent coatings is warranted.

Another area of application for ceramic coatings is in steam turbines. Component erosion by metallic-oxide scale in large steam turbines has led to the development of coatings that are erosion resistant at steam temperatures in the range of 538-566°C (1000-1050°F).[18] Tests conducted on stainless steel specimens with iron boride, chromium-enriched iron boride, and silicon-enriched iron boride coatings at 538°C and 500 ft/s for up to 100 h led to the conclusion that the enriched coatings were better than those without chromium and silicon, and the effect was attributed to an altered strain state, rather than to surface hardness, in the coating.

4 SUMMARY

Material degradation processes operative in coal-fired energy systems are outlined. The component requirements and material selections, which are wide ranging and system specific, are discussed briefly. Requirements for viable coatings for several applications are presented and factors that influence coating selection and application are enumerated. Discussions of coating types, coating methods, and several illustrations of performance of coatings in a number of applications are presented.

ACKNOWLEDGMENTS

This work was supported by the U.S. Department of Energy, Office of Fossil Energy, Advanced Research and Technology Development Materials Program, under Contract W-31-109-Eng-38.

REFERENCES

1. K. Natesan, J. Met., 1991, 43, 61.
2. K. Natesan, in Proc. Conf. Materials for Coal Gasification, W. T. Bakker, S. Dapkunas, and V. Hill (eds.), ASM International, Materials Park, OH, 1987, 51.
3. D. J. Baxter, High Temp. Technol., 1986, 4, 207.
4. E. C. Lewis, in Proc. Conf. Materials for Coal Gasification, W. T. Bakker, S. Dapkunas, and V. Hill (eds), ASM International, Materials Park, OH, 1987, 171.
5. P. A. Choquet, E. R. Naylor, and R. A. Rapp, Mater. Sci. Eng., 1989, A121, 413.
6. M. A. Harper and R. A. Rapp, Mater. Perform., 1991, 41.
7. K. Natesan and R. N. Johnson, Surface and Coatings Technol., 1990, 43/44, 821.
8. K. Natesan and W. F. Podolski, Argonne National Laboratory Report ANL-88-36, 1988.
9. J. Blough and W. T. Bakker, in Proc. Conf. Heat Resistant Materials, K. Natesan and D. J. Tillack (eds.), ASM International, Materials Park, OH, 1991, 567.
10. F. S. Kemp, Electric Power Research Institute Report EPRI AP-1369, 1980.
11. S. Shankar, D. E. Koenig, and L. E. Dardi, J. Met., 1981,33, 13.
12. K. Natesan, Mater. Sci. and Eng., 1987, 87, 99.
13. K. Natesan, Argonne National Laboratory, unpublished work.
14. K. Natesan and J.-H. Park, in Proc. Corrosion & Particle Erosion at High Temperatures, V. Srinivasan and K. Vedula (eds.), TMS, 1989, 49.
15. A. Bennett, Mater. Sci. and Technol., 1986, 2, 257.
16. T. N. Rhys-Jones and F. C. Toriz, High Temperature Technol., 1989, 7, 73.
17. D. J. Wortman, B. A. Nagaraj, and E. C. Duderstadt, Mater. Sci. and Eng., 1989, A121, 433.
18. T. H. McCloskey, Diffusion Coatings for Steam Turbine Components, Electric Power Research Institute Report EPRI TR-100208, 1991.

2.2.2
Advances in Zinc Coatings and Coated Steel Products

F.C. Porter

CONSULTANT TO ZINC DEVELOPMENT ASSOCIATION, UK

1 INTRODUCTION

The wide range of subject matter to be covered in this keynote address means that only a selected number of interesting developments can be included. Development has been very fast in the past two decades and the multitude of information has been gathered together in several places. The book on surface engineering practice based on the conference here in Newcastle in 1988 includes my survey on zinc coatings in surface engineering which gives a broad indication of the wide nature of the industry and the range of developments. A much more detailed survey covering all aspects of zinc is in the Zinc Handbook published in 1991, while I have just finished the text of a completely new edition of Zinc: Its Corrosion Resistance which will be published by ILZRO towards the end of 1992. These books have, of course, only been possible through great assistance by my colleagues in the zinc industry and particularly the facilities of the Zinc Development Association.

In this keynote lecture I have, therefore, selected three subjects to indicate the range of developments which makes zinc coated steel the subject of great development and interest at the present time. These are:

<u>(a) Hot dip galvanized coatings on fabricated steel products</u>

This industry takes approximately 30% of the two and a quarter million tonnes of zinc which are used for coating steel each year worldwide. The economic as well as the technical advantages of this coating, which produces a surface fully integrated metallurgically with the underlying steel, is being appreciated by more and more industries every year.

<u>(b) Coatings applied to steel sheet before fabrication</u>

The steel industry is acutely conscious of all added costs and, because most steel is covered by organic coatings for decoration, the steel industry is looking for a first coating which will give the necessary corrosion resistance and resistance against spread of rusting underneath the organic film. Zinc-base coatings remain the most economical but new and improved techniques of application as well as alloyed coatings are all finding their place in commerce.

(c) Thermally sprayed zinc coatings on concrete

Selected as an example of a coating chosen primarily for its electrical characteristics with corrosion resistance being a secondary factor.

2 HOT DIP GALVANIZING OF PRODUCTS

Steel Technology

The composition and properties of irons and steels - especially those of the surface - are fundamental to obtaining the most effective and economic hot dip galvanized coatings. The last twenty-five years has seen rapid growth in the production of steels by continuous casting. This requires special additions to the molten steel to obtain deoxidation. Aluminium, calcium or silicon-calcium are among the additions used. For hot rolled steels, the additions containing silicon often tend to be preferred by the steel industry. This can result in bulk concentrations in the steel of around 0.1% silicon; 0.04-0.12% bulk silicon gives excessive reaction in the galvanizing bath, the rate being linear with time and the coating having relatively poor coherence. It should be noted that many researchers (but not all) find enhanced concentrations of silicon at the surface so that work is still needed to clarify the surface composition and structure which should be avoided if possible.

To avoid excessive reaction, galvanizers initially operated at different temperatures, eg at 430°C to deal with susceptible steels instead of the usual 450-460°C. Specialized plants with ceramic baths operate in the upper parabolic range at 550-560°C which avoids the reactivity problem but can cause other problems and certainly costs more.

More general solutions were required, however. Early on, ILZRO proposed vanadium as an addition; this was technically satisfactory but gave operational problems. Then, in France, Noelle Dreulle and his colleagues developed the Polygalva process. The essential bath additions are lead (0.30-1.20%), aluminium (0.035-0.4%), magnesium (0.006-0.008%) and tin (0.03-0.15%). This process has been used commercially by some galvanizers, mainly in France and Belgium. It is a searching test for the efficiency of degreasing, pickling, prefluxing and flux drying operations. New high quality installations were usually needed. A few such installations have continued use of the process with success.

An easier solution was desired. Vieille Montagne - after reviewing the earlier ILZRO work - found other materials which could do the same job as vanadium and more easily. The main commercial development has become known as Technigalva and involves nickel in solution. Early work suggested 0.10-0.14% nickel in the bath was desirable. Industrial experience is that as little as 0.06-0.08% nickel may control the more reactive steels and has minimal effect on the less reactive steels, so that specification thickness minima can be maintained. In addition to eliminating the peak of fast reaction at around 0.08-0.1% silicon in the curve of coating mass against increasing silicon content, it also gives a good surface appearance. The slightly thinner coating achieved compensates for the higher cost of the alloy, even on non-reactive steels. The process is now used by many galvanizers in north-west Europe and in Australia. Major benefits are that the coating performance is unchanged, thickness for thickness, and that no changes to existing plant are needed to introduce the process.

Above about 0.3-0.4% silicon in the steel, the Technigalva process has less effect. However the thick and coherent alloy coatings that form on such steels are often slightly thinner than if nickel were not present and give good service performance. Thick coatings are needed on some products, eg wall ties at 960 g/m^2 and such relatively high silicon contents achieve this.

Hot dip galvanizing did operate largely unchanged for 150 years. Now, in addition to the alloy development already mentioned, the past twenty-five years have seen new techniques such as the 'outside only' galvanizing of gas bottles and steady improvements in process control to ensure a still more uniform product. It is now much more common for degreasing to be part of the standard operating procedure in order to avoid contaminating later baths in the sequence. Pickling in inhibited acid and controlled immersion times both give more reproducible surfaces and ensure that any limitation in the steels used are not aggravated. Environmental controls have been steadily introduced - both to improve conditions in the workplace and to control exhausts leaving the works.

The average size of a galvanizing plant for products has been increasing steadily. Countries such as Great Britain now have an average yearly throughput of over 5,000 tonnes per plant -fifty years ago the average was less than 2,000 tonnes per plant. Size of plant on average will tend to increase in future because it will become uneconomic for small plants to operate with the rigid environmental controls that are now necessary. About 30% of galvanizing is currently by fabricators for their own products but such specialized product lines are more likely to be integrated with bigger general galvanizing plants in future in order to utilize extensive environmental control equipment to the greatest extent. Although the fume from galvanizing pots is harmless, it is visual and the emotional reaction - which is a feature of our modern society - is likely to mean that most galvanizing plants will need to contain such fume.

Standards and Certification

The last twenty-five years has seen progress in agreeing an international standard for the galvanized coatings on fabricated products. The first ISO 1461 published over 15 years ago could not give a thickness for the coating on steel between 2 and 5mm thick because of the lack of agreement between countries. Now, at long last, in 1992 a document fully agreed in both CEN and ISO committees - after notable concessions by almost every country- is on the point of being issued. The user industries and the galvanizers should rightly congratulate themselves on this achievement. The alloy developments mentioned above are not directly covered in the standard - they are a matter for individual galvanizers expertise. It is only necessary for the product to meet the requirements of the standard. In the next year or two this document, together with the ISO/CEN document on thermal spraying, will be the base for a guidance document on the protection of steel in structures (required as an adjunct to Eurocode 3) for use throughout the European Community.

The industry is also able to congratulate itself on the increasing introduction of certification schemes and of self-certification based on the precepts of the ISO 9000/EN 29000 series of documents which are paralleled in national standards such as BS 5750.

Returning to the galvanizing of products standard, the unique feature of galvanized coatings from the user's viewpoint is that the nature and thickness of the ferrous material on which they are formed determines the specification thicknesses. Although accepted by the user, this practice does give rise to problems such as thinner coatings on fasteners than on the heavy structural items which they join. It will be of benefit to the galvanizing industry if they can offer equal corrosion resistance on all parts of the structure, irrespective of material type or thickness. Users require that threaded parts of course have a maximum limit on thickness of coating if they are to do their work as fasteners and, therefore, the most likely development is to use alloy coatings on fasteners, wire and sheet to develop corrosion resistance at least twice that of a traditional galvanized coating.

For this reason, more research work is needed on the zinc-5% aluminium (Galfan) type of composition to make it suitable for general galvanizing. The fast-cooled versions of Galfan that are available on continuously coated sheet and wire products are significantly superior to the earlier slow-cooled versions and definitely offer a factor of at least twice the corrosion resistance for the same thickness of coating.

3 ZINC AND ZINC ALLOY SHEETS

General considerations

The borderline between the galvanizing of fabricated products and the continuous processing of sheet, wire and tube is becoming more blurred with time. Continuously galvanized sheet may be as thick as 7mm and, when formed, can compete with many products galvanized after manufacture - especially where the extra durability of the galvanized-after-manufacture product is not essential. With wire, many products can either be made from pregalvanized wire or can be galvanized after fabrication, eg mesh.

Alloy development

Zinc is mostly widely used because of its low cost and widespread availability. Aluminium hot dipped coatings are more expensive than zinc hot dipped coatings and are, therefore, mainly restricted to uses where their higher temperature corrosion resistance is required and also for specialized building applications.

The use of zinc-aluminium alloys to combine the best properties of both zinc and aluminium has been an attractive proposition for many years. The zinc-55% aluminium-1.5% silicon alloy coating gives a product that has several times the corrosion resistance of pure zinc but, in achieving this, loses some of the other desirable properties of the coating -including limitation of the degree of cathodic protection.

Ten years ago, therefore, research started on a composition with only about 5% aluminium. The product - Galfan - which has now been commercially available on wire and sheet for some years, fully retains the cathodic edge protection and paintability of the conventional galvanized coating. It has substantially enhanced corrosion resistance and improved formability - even better than galvanized sheet. Long term atmospheric testing in industrial, marine and rural atmospheres shows corrosion resistance over twice as great as conventional zinc coatings in the unformed state and still better results, both painted and unpainted, when in the formed condition because of the restricted formation of coating cracks in the deformed area.

At present, about 45 firms are licenced to produce Galfan coated steel with over 20 lines in commercial production of sheet, wire and tube. Licensees are spread worldwide to cover all six inhabited continents. Major commercial production of Galfan now occurs in Europe, Japan and North America. Applications are as varied as for galvanized sheet steel and there are long term cost savings. ASTM specifications already exist and European and other national specifications are being developed. The prestige of the coating has also led it to be used in such notable projects as the cables for Le Nuage, the new Arch de la Defense in Paris which now forms a twin attraction to the Arch de Triomphe.

While it is difficult to foresee the future, the special and different properties of the three materials (zinc, zinc-5% aluminium and zinc-55% aluminium) may lead to the total market expanding substantially with each coating taking about an equal share of the market. Hot dip coatings with aluminium contents between 5 and 55% have not had commercial success.

The most exciting growth area for hot dip zinc-coated steel sheet has been cars such as the Audi range. Other manufacturers have used galvannealed sheet - in-line heat treatment converts all the zinc coating to an iron-zinc alloy which has easier welding and paintability characteristics.

Alloy Coatings by Other Techniques

Electroplating permits a wide range of zinc alloys to be deposited. One of the most successful has been that containing about 12% nickel which has performed particularly well in accelerated tests although the degree of improvement to be obtained in practice is not yet quantified. Most alloy additions are only, however, a fraction of one percent. Nickel, iron, cobalt and chromium are amongst the alloy additions being used and the study of their development forms an ever-expanding subject which is extensivly covered in the three-yearly zinc coated sheet conferences and in the Galvatech conferences, Tokyo in 1989 and Amsterdam in September 1992.

The pretreatment (pickling) and electroplating process can, however, introduce hydrogen embrittlement into high strength steels and consequently non-electrolytic coating methods are preferred for such steels. Thus, mechanical plating rivals electroplating especially for high strength steel components such as bolts.

4 ZINC SPRAYING OF CONCRETE BRIDGES

An indirect use of zinc for corrosion protection is to provide a low-resistance path for impressed current protection of concrete bridge structures with uncoated steel rebar. Electrochemical protection of steel rebar in concrete was developed as a repair technique but is now being promoted also for new structures. Impressed current is generated from inert anodes outside the concrete and this applied current stops the rebar from rusting.

The primary anode is usually brass or copper. The secondary anode is either (i) conductive paints or polymers (ii) titanium wire mesh or (iii) zinc thermal spray - although one bridge used zinc sheet scrap from blanking one cent pieces! Development has been in that order with zinc spray trials starting in 1982. The zinc is sprayed directly onto the heated outer concrete surface of the bridge. Direct contact between the zinc spray and the steel rebar has to be avoided as otherwise the zinc would be prematurely consumed. Exceptionally, Florida DOT sometimes use the zinc as a sacrificial anode (eg, in 1988 on spalled areas with rebar exposed on the Niles Channel Bridge; on the Julia Tuttle Bridge, Miami, 1989 and the McCormick Bridge, Jacksonville) as they often have concrete permanently wet and salt-saturated.

Californian DOT 1982 evaluated Al, Al-Si, 80%Ni-20%Al, zinc, stainless steel, sprayed Babitt (Sn-Sb-Cu), C-filled latex paint, etc. They considered conductivity, consumption, adhesion, weather resistance, aesthetics, cost, availability as wire and compatibility with salt-impregnated concrete. Babitt or zinc were best. The initial trials were on selected areas of:

(i) Richmond San Rafael Bridge, San Francisco - two columns 20m^2, 225 μm zinc - in good condition after 7 years.

(ii) East Camino Undercarriage, Placeville - Eastbound with zinc arc-sprayed on deck with striped pattern to ensure concrete bond to asphalt overlay 12.5cm zinc 12.5cm gap 300m^2 225 μm zinc Westbound - thermally sprayed 500 μm zinc on soffit, 150m^2.

Three hundred USA bridges now have cathodic protection and several more are done each year. Also, recently, 100m^2 of a parking garage has been done using zinc. The economics are reasonably encouraging.

5 CONCLUSIONS

The last twenty-five years has seen a sharp increase in the competition between rival materials and finishes. Steel itself has a long standing fight with concrete, with plastics and with aluminium, for markets. The availability of a wide range of corrosion resistant and coloured steels can greatly assist steels in retaining and, indeed, obtaining markets. Zinc in itself cannot provide colour other than grey but its presence underneath paint or plastics is essential if the latter are to last. Of the zinc coating, hot dip galvanizing is the most important but has to compete with other finishes. Metal spraying with zinc, with aluminium and with alloys can be cheaper than galvanizing on heavy sections. Mechanical coating can be cheaper on some small parts and has claims for technical advantages in relation to embrittlement with the very high strength steels. Sherardizing is particularly effective for uniform coating on threaded work and is currently the choice for many threads up to about 6mm diameter -it is certainly a much better coating for these parts than electroplating.

REFERENCES

1. F. C. Porter, 'Zinc Coatings in Surface Engineering', Surface Engineering Practice, Ellis Horwood, 1990, 306-314.

2. F. C. Porter, 'Zinc Handbook - Properties, Processing and Use in Design', Marcel Dekker Inc, New York, Basel, Hong Kong, 1991, 629pp.

3. ILZRO, 'Zinc: Its Corrosion Resistance' 3rd Edition. In the press.

2.2.3
A New Technique for Small Area Corrosion Testing for the Analysis of Variable Composition Laser Melt Tracks

M.A. McMahon, K.G. Watkins, C. Sexton,[1] and W.M. Steen[1]

DEPARTMENT OF MATERIALS SCIENCE AND ENGINEERING, UNIVERSITY OF LIVERPOOL, PO BOX 147, LIVERPOOL L69 3BX, UK

[1] DEPARTMENT OF MECHANICAL ENGINEERING, UNIVERSITY OF LIVERPOOL, PO BOX 147, LIVERPOOL L69 3BX, UK

1 INTRODUCTION

Bulk solution corrosion testing usually involves taking a sample and placing it in a cell as one of the electrodes. This method is dependent on having either small samples or relatively large corrosion cells. Micro-corrosion tests have been developed successfully to analyse very small samples with the advantage of being able to produce a more even potential distribution as well as the possibility of achieving high current densities because of the extremely small areas involved[1]. These types of tests also have the disadvantage of requiring intricate cells that are hard to observe and are usually very expensive.

For large objects corrosion tests are usually carried out by an exposure test or immersion test which has to be monitored over long periods of time[2]. These methods, such as exposing large plates of metal to an industrial atmosphere for example, are only very general in the information that they provide. They only give, at the time of testing, a limited indication as to what might be happening at a specific area where the local surface properties may for some reason be different to that of the bulk surface properties. These properties can subsequently be observed but only by testing smaller individual sectioned pieces of the overall object. It is then desirable to be able to analyse the corrosion properties of small areas without having to make a very small sample which involves destroying the actual object. In the case of this work the particular bulk samples are variable composition laser melt tracks[3,4]. The tracks are deposited onto a mild steel substrate via a three hopper, single feed tube blown powder technique. A schematic diagram of the technique is shown in Figure 1. Microprocessor control of the three hoppers allows laser melt tracks to be prepared with the composition varing along the entire length. In this way metered sections of a complete ternary alloy system can be produced on a single steel plate.

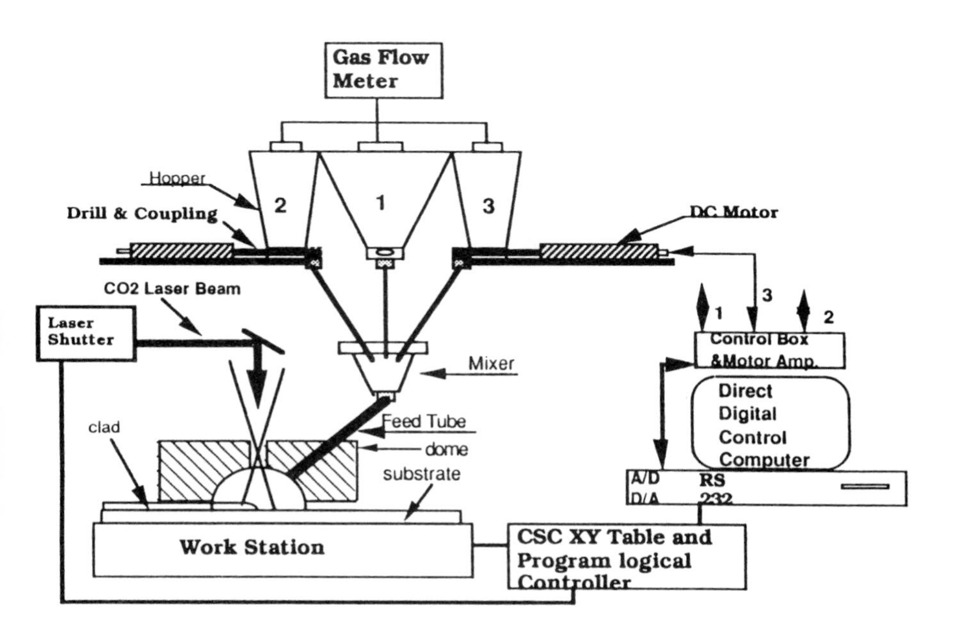

Figure 1 Triple hopper single feed laser cladding technique

With the eventual aim of being able to test the above mentioned melt tracks, a bridge type probe has been developed. The probe allows electrochemical analysis by delivering a drop of electrolyte and holding it in position over the area to be tested. By this method the corrosion properties of the sample can be measured and expressed in terms of the electrochemical parameters; polarization resistance(Rp), anodic and cathodic Tafel constants(b_a and b_c respectively), which can all be obtained graphically. The corrosion current(I_{corr}) can then also be calculated by applying the Stern-Geary equation, by extrapolation[5,6], or by techniques that fit the polarization data to the Wagner-Traud equations, such as the POLCURR program[7,8]. The POLCURR program (which gives values for Rp, I_{corr}, b_a and b_c) is important when polarization measurements need to be restricted to avoid contamination of the electrolyte drop by corrosion products. In this case all four parameters can be found by carrying out a single polarization test near to the free corrosion potential.

In order to show that the probe can be used to carry out polarization tests in a drop near to the free corrosion potential(Er) pure zinc foil was used as a standard material. Tests were aimed at investigating whether the results obtained using the salt drop probe are similar to results from bulk solution tests.

2 EXPERIMENTAL PROCEDURE

A schematic diagram of the probe and the experimental set up is shown in Figure 2. A very slight positive flow through the probe ensured that there was a fresh supply of electrolyte for each experiment.

The zinc used in this work was commercially pure zinc foil. The surface of the working electrode was prepared by grinding with silicon carbide paper to a 1200 grit finish. It was then washed and degreased using Genklene detergent and dried with alcohol. This was followed by abrading the surface for a second time with 1200 grit paper and washing in distilled water in an ultrasonic bath.

Polarization of the samples was carried out with a Thompson Electrochem Autostat, a microprocessor controlled potentiostat. A saturated calomel reference electrode(SCE) was used and a platinum wire inside the probe was used as the auxiliary electrode. Potential sweeps from -20mV to +20mV were carried out using a scan rate of 60mV/min.

The electrolyte used was 1N NaCl at pH4. This was prepared with distilled water and AnalaR grade chemicals. The pH was adjusted with 1M hydrochloric acid.

All areas were masked off using Lacomit varnish.

3 RESULTS AND DICCUSSION

Typical results for a bulk solution test and for the probe are shown in Figures 3(a) and 3(b), respectively. The results are similar in most respects but when the

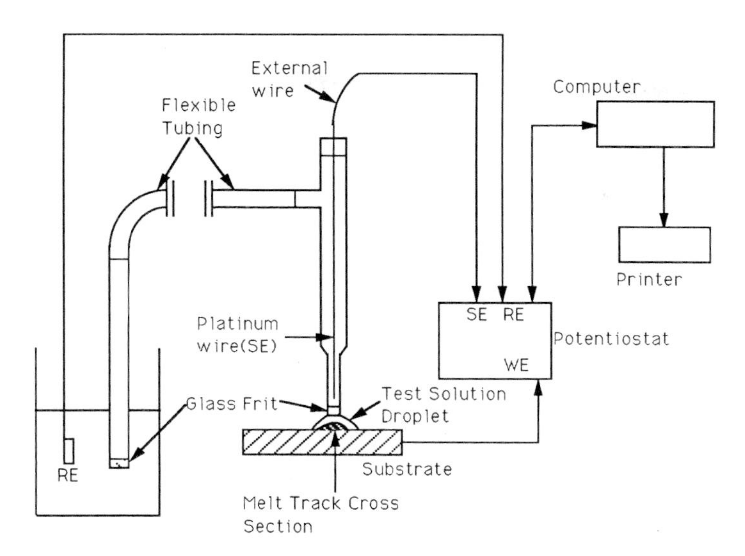

Figure 2 Salt bridge probe and experimental setup

currents are compared the anodic current for the probe seems to have been greatly reduced. It can be seen that the cathodic current in the probe test is not that dissimilar to that of the bulk solution test. However, the anodic current is different.

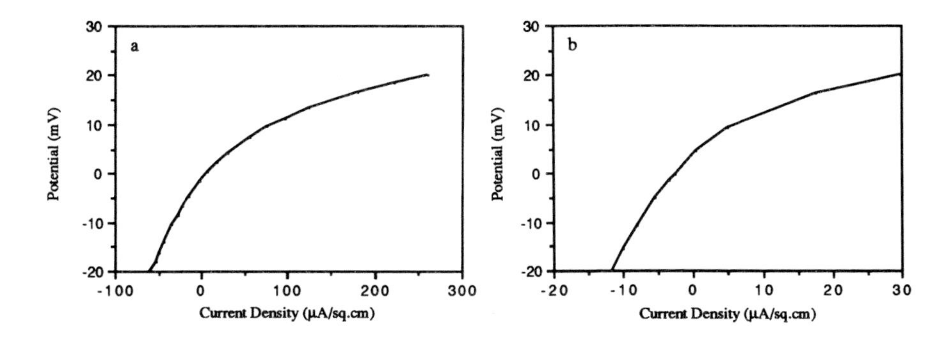

Figure 3 Polarization curves for zinc (a) bulk solution test, (b) probe drop test

The probable cause for the difference is differential aeration within the drop at the metal surface. The oxygen concentration is thought to be greater for the drop than for bulk solutions because of the greater ability of oxygen to diffuse to the surface of the metal. In addition, oxygen diffuses more rapidly to the metal at the edge of the drop than at the centre and this causes differential aeration to occur. The outcome of this is that as the polarization tests proceed and the potential increases anodically, the metal at the circumference of the drop is forced to have a more negative potential than that of the central region. This results in the overall anodic current being suppressed sufficiently to alter the form of the polarization curve.

If the triple boundary interface between the air, solution, and metal (see Figure 5(a)) is eliminated from the experiment then the results correspond. The curves shown in Figure 4 show the effect of decreasing the effective length of the triple boundary on the anodic

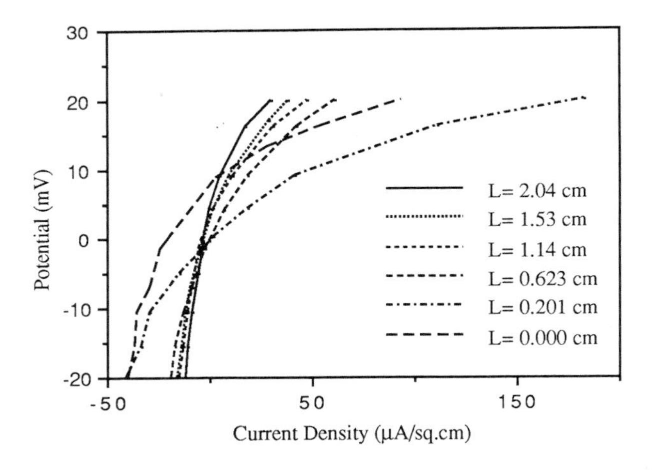

Figure 4 The effect of decreasing the length of the triple boundary on polarization measurements

current. This was achieved by masking off areas under the drop in the sequence shown schematically in Figure 5(b) and carrying out further polarization tests. Now comparing the curve for the sixth test with the curve for the original bulk solution test it is clear that the currents are more similar. There is a slight difference in the current density between the two tests giving a higher value for the probe test. This is thought to be due to the effect of the tested areas being different in size and shape[9]. These results also indicate that it is not always possible to assume that non-uniformity of current density as a consequence of an 'edge effect' is negligible. Pontinha and Ferreira[10] have found from a.c. impedance spectroscopy measurements of zinc that with an area of 0.785 cm^2 pitting occurred at the centre, but with a smaller area of only 0.049 cm^2 no pitting was observed.It may be that for accurate corrosion testing smaller areas are to be preferred to avoid 'edge effects' which cause non-uniformity in the current density.

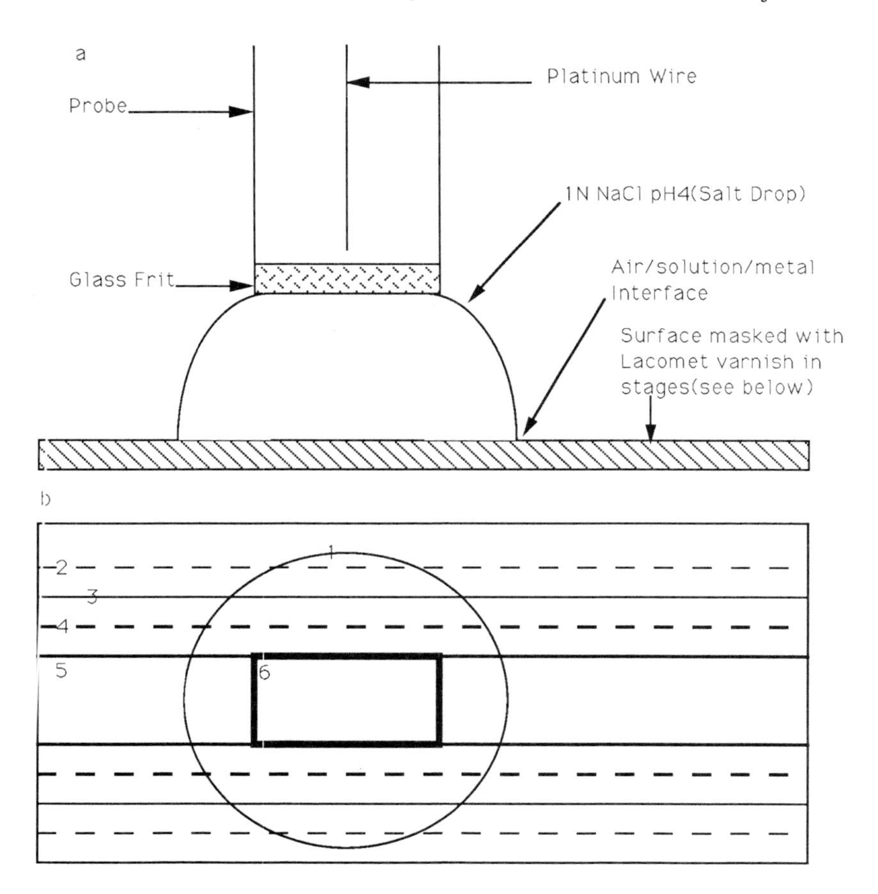

Figure 5 (a) Electrolyte drop in position at the end of the probe, (b) masking sequence 1=100% interface, 6= no interface

Another way to view the results is shown in Figure 6. The apparent polarization resistance(Rp_A), measured as Rp for the different tests, has been plotted as a function of the length of the triple boundary interface. It would seem that there is a linear relationship between them. The value for Rp_A at the intersection of the vertical axis is akin to values of Rp that are regularly obtained for zinc in bulk solution tests[11].

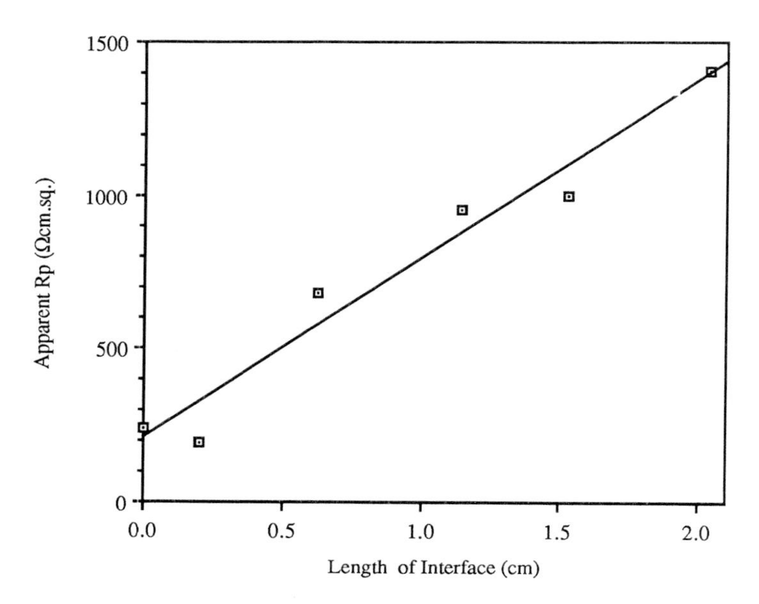

Figure 6 Effect of decreasing the length of the triple boundary on the value of Apparent Rp

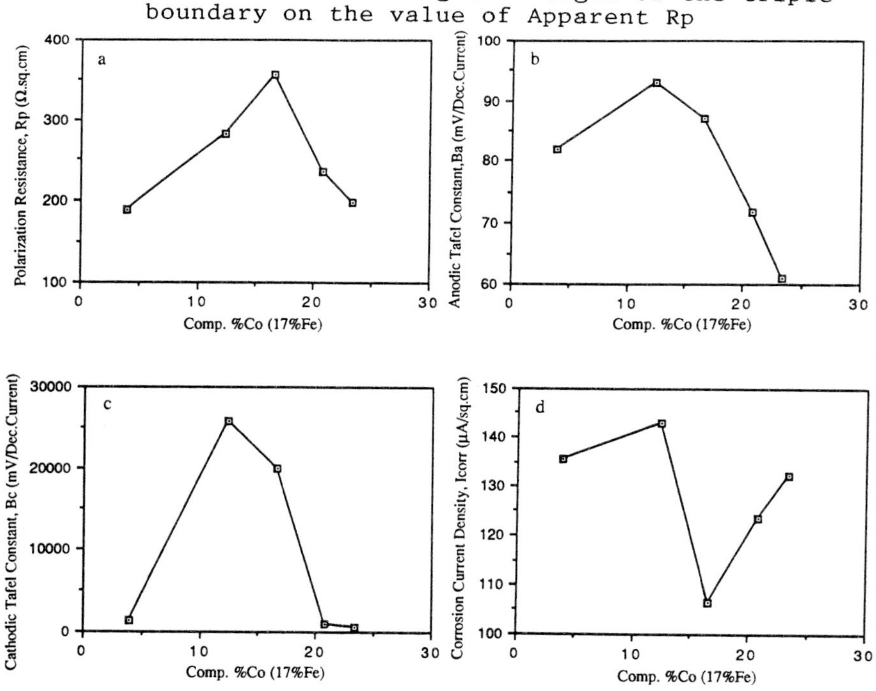

Figure 7 Results obtained from the Co-Fe-Ni system plotted against composition; (a) Rp, (b) b_a, (c) b_c, (d) I_{corr}

Some investigations have been carried out using the probe for alloys taken from a variable composition laser melt track for the cobalt-iron-nickel system. In this case the electrolyte used was 1N H_2SO_4 as this system appeared to be extremely resistant to corrosion in 1N NaCl at pH4. The results obtained from the POLCURR program, for one of the melt tracks, are shown in Figure 7. Rp, b_a, b_c, and I_{corr} respectively are plotted against composition. The probe was also used to monitor the change in Er with time, see Figure 8. This type of information can define the sort of corrosion as either general corrosion, general passivation, or pitting corrosion[12]. The curve in Figure 8 suggests that this particular alloy undergoes pitting corrosion.

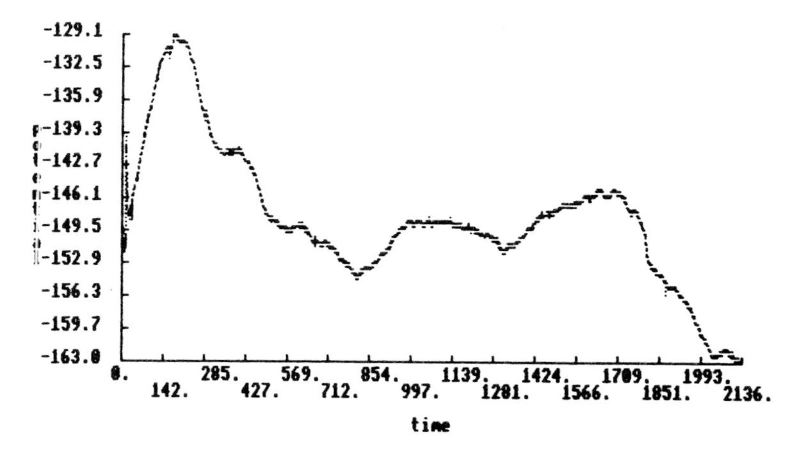

Figure 8 Variation of rest potential with time

Anodic polarization sweeps upto 300 mV were also carried out and the curves obtained for the system are shown in Figures 9 (a) to (f). The curves show the effect of difference in alloy composition on the anodic reaction. The reaction indicated by these curves is probably for sulphate formation since passivation clearly has not occurred. It would seem that increasing the cobalt content suppresses this reaction. Full anodic sweeps could also be carried out to investigate further reactions such as passivation and the trans-passive reaction.

4 CONCLUSION

1. A salt bridge probe has been developed which enables polarization measurements to be carried out in a small drop of electrolyte. The technique has been shown to work if differential aeration in the drop is eliminated.

2. Polarization resistance measurements were carried out

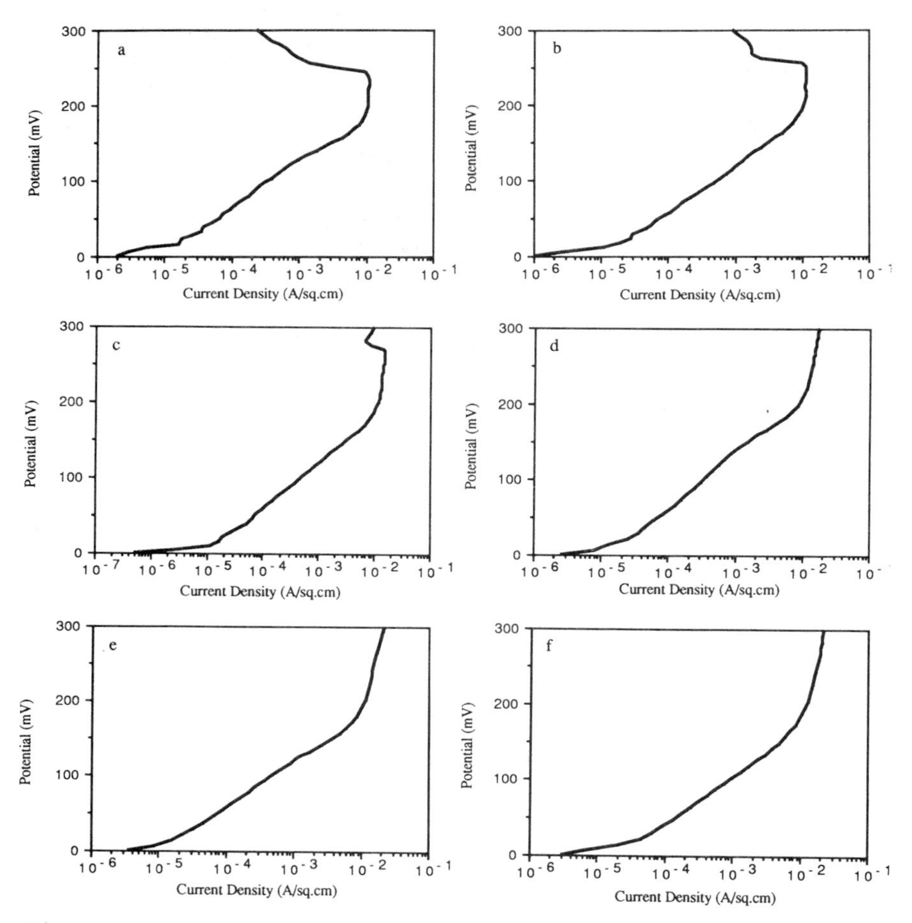

Figure 9 Anodic polarization curves for Co-Ni-10Fe laser melt track with variation in cobalt and nickel; (a) 4Co, (b) 7Co, (c) 10Co, (d) 16Co, (e) 19Co, (f) 26Co; values in wt%

on zinc and the values obtained for Rp were akin to those for bulk solution tests.

3. The results obtained indicate that using working electrodes of small areas ($<0.05cm^2$) possibly gives greater accuracy when measuring Rp.

4. The probe has been successfully used to analyse laser melt tracks from the cobalt-iron-nickel ternary system and some of the results have been presented here. Furthermore, it has been shown that the probe can carry out polarization over quite large potential shifts and is not restricted to potential sweeps near to Er, provided the electrolyte composition remains unaltered as a result of the anodic reaction.

ACKNOWLEDGEMENTS

The authors would like to acknowledge the support of the EC SCIENCE program (Project PL90100136).

REFERENCES

1. K.Wkiel and J.Osteryoung, J.Electrochem.Soc., 1988, 135,No.8, 1915.
2. F.Mansfeld."Evaluation of Electrochemical Techniques for the Monitoring of Atmospheric Corrosion Phenomena", Electrochem.Corr.Test., Am.Soc.Testing and Mats., 1981, 215.
3. W.M.Steen and K.G.Watkins, Forthcoming at 3rd Int. Symposium on High Temperature Corrosion and Protection of Metals, 25-29 May, 1992, Les Embiez, France.
4. J.Y.Jeng, B.Quayle, P.J.Modern and W.M.Steen, Forthcoming at LAMP'92, Japan.
5. M.Stern and A.I.Geary, J.Electrochem.Soc., 1957, 104, No.1, 56.
6. M.Stern, Corr.NACE., 1958, 14, No.9,60.
7. F.Mansfeld, Adv. in Corr. Sci. and Tech., 1976, 6, 163.
8. S.M.Gerchakov, L.R.Udey and F.Mansfeld, Corr. NACE, 1981, 37, No.12, 696.
9. D.Remppel and H.E.Exner, J. Electrochem. Soc., 1991, 138, No.2, 379.
10. Private Communication
11. K.G.Watkins, R.D.Jones and K.M.Lo, Materials Letters, 1989, 8, 21.
12. T.P.Hoar and D.C.Meers, Proc. Roy. Soc.(A), 1966, 294, 486.

2.2.4
Corrosion Resistant Coatings of Aluminium Bronzes

T. Diakov, L. Georgiev, and E. Georgieva

BULGARIAN ACADEMY OF SCIENCES, INSTITUTE OF ELECTRONICS, 72 TRAKIA BOULEVARD, 1784 SOFIA, BULGARIA

1 INTRODUCTION

In recent years the method of electron beam evaporation of metals has been actively used for the fabrication of corrosion resistant coatings based on binary Al-Cu alloys.

Basically, this is a physical process of vacuum evaporation whereby a vapour flow is created using the kinetic energy of the accelerated electrons directed towards the evaporators. The coatings obtained by this approach are of high purity and their structure and properties can be controlled within a wide range[1]. Moreover, from a technological point of view, such coatings can be formed with highly reproducible results.

2 EXPERIMENTAL

The experiments involved in this study were carried out on the equipment shown in Figure 1. The electron beam optical system is placed horizontally. It has an independent anode and separate devices which can both focus and deflect the electron beam. The deflection angle is 90^0 and there is also a possibility of scanning the beam over the two targets. A controllable and stabilized 10kW power supply is used. The process of evaporation takes place in a vacuum chamber. The vacuum system maintains a working vacuum of 10^{-2}Pa. Inside the chamber there is a bulk water cooled copper crucible with two cavities each of 35 mm diameter; these accommodate the starting materials, namely Al of 99% purity and refined Cu. The substrates to be coated are placed at a distance of 200 mm from the crucible. They are made of low carbon steel of the following chemical composition: 0.09-0.15% C; 0.3-0.5%

Mn; 0.055% S; 0.05% P and trace Si. Prior to coating, the substrates were subjected to a mechanical treatment in order to enhance the adhesion of the deposited layers, and then polished, degreased and heated to a temperature of 430^{0}C in vacuum by means of an electrical heater. Before beginning the process the substrates are screened with a shutter which is removed upon achieving the necessary working regime.

The electron beam facilities for evaporation of metals are equipped with control and measurement units which are mandatory for the coating process to run properly. In our case the metals are evaporated separately while their condensation on the samples takes place simultaneously. With the aid of a preliminary scanning programme it is possible to deflect the electron beam from one evaporator to another and vice versa.

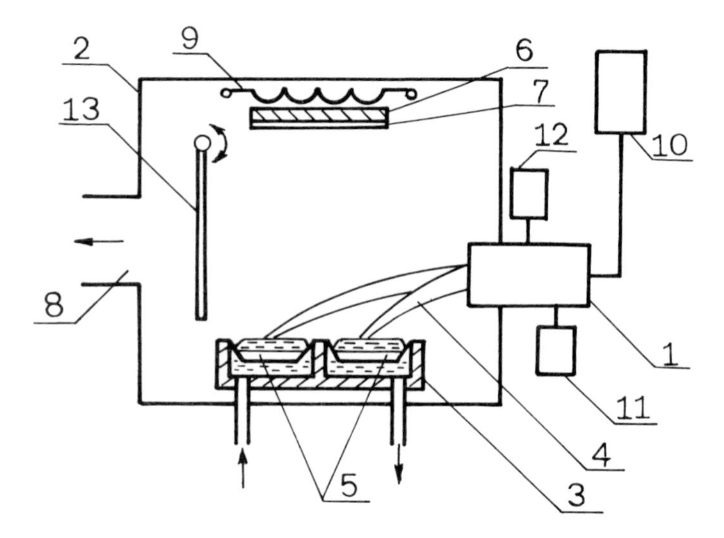

Figure 1 Schematic of the experimental equipment
1. electron beam system; 2. vacuum chamber; 3. crucible; 4. electron beam; 5. evaporated materials; 6. substrate; 7. coating; 8. to vacuum system; 9. heater; 10. high voltage source; 11. focusing device; 12. deflection device; 13. shutter

By taking into account the specific rate of evaporation shown in Figure 2, the duration of the beam influence is chosen in such a way as to reach the appropriate temperature ranges at the surface of the materials being evaporated. Changing the above mentioned duration enables one to obtain coatings of different composition, for example, to attain a coating of 10% Al, the ratio of times is ca. 6:1.

3 RESULTS AND DISCUSSION

Some typical regimes for forming coatings of 100-350 nm thickness are given in Table 1. The structure of the resulting coatings was determined using a JEOL/35 CF scanning electron microscope with an X-ray microanalyser. The analysis revealed an equilibrium microstructure similar to that of recrystallised metals and alloys. The corrosion resistant properties of the coatings are associated with the oxidation of the surface Al and subsequent formation of a thin oxide layer. When Cu was added the temperature durability of the coatings was found to increase. In the case of a steel sample with a Cu-30% Al coating obtained by this method and placed in air at 700^{0}C for several days, no rust stains were discernible.

Table 1 Regimes of coating formation

I mA	U kV	P kW	D mm	p W/cm^2	v nm/s
90	26	2.3	3	7.6x10^3	1.25
100	26	2.6	3	8.6x10^3	0.9
120	26	3.1	3	1.0x10^4	0.95
150	26	3.9	4	7.8x10^3	0.85

The protective properties of Al bronzes in various aggressive ambients such as moisture, salt water, acid and alkaline media, depend strongly on their porosity. Pores are known to be controllable structural features of the coatings. The temperature of deposition and the rate of condensation both have a profound effect on the layer porosity. Using a transmission optical microscope of magnification greater than 250X the average number of pores was determined over a unit area of an Al-10% Cu coating deposited at room temperature and at 300^{0}C, with all the rest of the process conditions being kept the same.

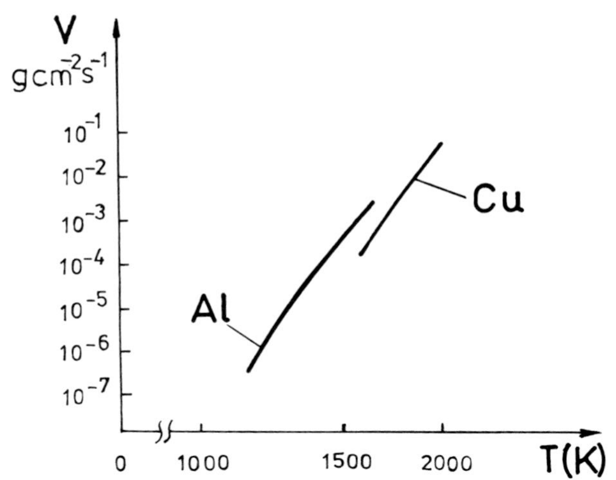

Figure 2 Specific rate of evaporation for Al and Cu as a function of temperature

The results showed that the average number of pores decreased from 5600 cm^{-2} at a temperature of deposition of 27^0C, to 1700 cm^{-2} for deposition at 300^0C. Other parameters which influence the porosity are the degree of substrate roughness, initial temperature and layer thickness. The degree of porosity can be estimated on the basis of the data given in Reference 2.

The samples were tested under different conditions, namely (i) 6 min. treatment in a 40% aqueous solution of KOH, (ii) 48 h treatment in a 3% solution of NaCl, and (iii) air ambient. Slight changes in the coated samples became only discernible at the end of the above time intervals while bare uncoated samples showed active etching accompanied by the appearance of rusty spots.

For better quantitative assessment of porosity, several samples were prepared on transparent substrates together with the original substrates, and were subsequently examined using a quick photoelectrical method. Calibrating the photocurrent generated by the transmitted light enables an evaluation to be made of both the amount and size of the pores over a unit surface of the coating after treatment in an aggressive ambient. The average values of the photocurrents prior to and after a 6 min. immersion in 40% KOH were measured for samples coated with Al-10% Cu at room temperature and at 300^0C. The results showed a 3 orders of magnitude increase in the photocurrent for the room temperature samples, and an increase by a factor of 30 for the samples prepared at 300°C.

The adhesion force was determined by means of the two most exploited methods: the scotch-tape peel-off test and the peeling-upon-folding measurement. The first test was applied to all the coatings for preliminary evaluation of adhesion. The second method showed that best adhesion was achieved in the case of mirror-polished samples of 90% Cu and 10% Al composition coated at an initial substrate temperature of 300-350°C. Some of these samples were capable of withstanding more than 100 foldings without any observable cracking in the coating.

The microhardness of the samples was determined on the basis of the Vickers' method employing a PMT-3 instrument. A study was made of the variation of microhardness by increasing the percentage of Al in the coating from 10 to 40%, and an increase in the microhardness by a factor of 7 was observed. A wide microhardness range was established which is presumably due to the various temperature regimes of sample coating.

4 CONCLUSIONS

1. The use of electron beam evaporation is proposed which permits the fabrication of Al-Cu coatings within a wide range of composition.

2. The analyses and measurements carried out confirm that the resulting coatings possess pronounced corrosion resistant properties and good chemo-physical characteristics.

REFERENCES

1. Z. Shiler, U. Gaizig and Z. Pantzer, "Electron Beam Technology," Energia, Moskva, 1980.
2. I. Roih, L. Koltunova and S. Fedosov, "Application of Protective Coatings in Vacuum," Mashinostroene, Moskva, 1976.

2.2.5
Chemical and Electrochemical Surface Modification Techniques for the Improvement of Copper Corrosion Resistance

S.B. Adeloju and Y.Y. Duan

ELECTROCHEMICAL AND CORROSION TECHNOLOGY RESEARCH GROUP, DEPARTMENT OF CHEMISTRY, UNIVERSITY OF WESTERN SYDNEY, NEPEAN, PO BOX 10, KINGSWOOD, NEW SOUTH WALES 2747, AUSTRALIA

1 INTRODUCTION

Copper and its alloys have been used extensively for several decades in the transportation of water for domestic and industrial purposes. Current areas of application of these materials include steam condensers, heat exchangers, distribution systems for potable and industrial water, where they have demonstrated reliable performance and extended the life of such systems[1-4]. The excellent performance of these systems is attributed to the high corrosion resistance of copper and its alloys in various aqueous environments[4,5]. However, the metal can corrode in some environments due to the presence of some aggressive substances such as chloride, which is often associated with the pitting corrosion of copper.

Several workers[6-9] have investigated the corrosion behaviour of copper under various solution conditions. Some of these workers studied the influence of bicarbonate, chloride, sulphate, pH, residual chlorine and buffer capacity. Steve et al.[10] evaluated the relationship between copper plumbing corrosion and variations in water quality in the Pacific Northwest. These workers found that the oxide layers on aged copper surface provided a substantial protection under the prevalent condition. In comparison with clean copper surfaces, the oxide film on the aged surfaces reduced the corrosion rate by about 50 per cent.

Some workers have also studied the anodic behaviour of copper in neutral, alkaline and aerated solutions of different ionic composition by electrochemical methods[11-14]. In particular, the nature of the anodic oxide film formed on copper surfaces has been studied with various electrochemical and surface analytical techniques[1,12-14,15-20]. These various studies have clearly demonstrated the ability to form various oxides and basic copper salts on copper surfaces by use of anodic polarisation in aqueous media, but these findings have rarely been exploited for the improvement of the corrosion resistance of the metal. Yet, the development of a systematic and simplified approach for the deliberate formation of the passive film(s) on copper surfaces may have some potential in improving its corrosion resistance. In particular, the prior modification of copper surfaces by use of a simple chemical and/or electrochemical treatment under specific controlled conditions could serve as a basis for improving its corrosion resistance in aqueous media. For example, under feasible conditions, it will be possible to completely stifle the corrosion of a metal by the prior formation of a specific stable protective layer.

In this study, the utilisation of chemical and electrochemical surface modification approaches for the improvement of the corrosion resistance of copper in aqueous media is explored. This involves the development of strategies for the deliberate and selective formation of protective film component such as copper (I) oxide, or copper (II) oxide, or a mixed oxide layer under carefully controlled conditions. Scanning electron microscopy and X-ray diffraction analysis will be used to examine the surface modification process, identify the composition and morphological arrangement of the films on copper surfaces. The extent of the improvement of corrosion resistance achieved by the protective layers will be investigated by weight-loss, and potentiostatic polarisation methods.

2 EXPERIMENTAL

Modification of Copper Surfaces

A copper sheet of 0.5 mm in thickness (99.9% purity with other minor constituents) was cut to a coupon size of 50 mm x 25 mm The specimens were further prepared by etching the surface in 3.0 M nitric acid, rinsing with Milli-Q water, polishing with 400 and 600 grade wet or dry waterproof silicon carbide sandpaper, rinsing with acetone, washing again with Milli-Q water and finally air dried. The specimens were then immersed in sodium hydroxide solution (1M) for a 5-day period to form the oxide layer.

Weight Loss Method

Following the film formation, the specimens were taken out of the solution, rinsed with Milli-Q water, placed on a tissue paper and left to dry. The specimens were then weighed (W_o) before being immersed in the different solutions, as given in Table 1. Duplicate measurements were made for each specimen. To determine the weight change the specimens were taken out, cleaned, dried and weighed as previous (W_t). The weight loss or weight gain was determined from the change in the weight of the specimen by the following formula:

$$\% \text{ Weight Change} = (W_o - W_t) / W_o \cdot 100$$

Table 1 The solution and surface conditions used for the study of the corrosion resistance of modified copper surfaces

Group	Specimens	Exposure Solutions		
		HCO_3^- mg/l	Cl^- mg/l	CO_3^{2-} mg/l
G 1	Copper + Cu_2O/CuO film	31	18	0.06
G 2	Copper + Cu_2O/CuO film	31	36	0.06
G 3	Copper + Cu_2O/CuO film	31	177	0.06
G 4	Copper + Cu_2O/CuO film	61	31	0.06
G 5	Copper + Cu_2O/CuO film	305	36	0.06
G 6	Copper + Cu_2O/CuO film		36	
G 7	Copper + Cu_2O/CuO film	31		0.06
G 8	Copper + Cu_2O/CuO film	61		0.06
G 9*	Copper + Cu_2O/CuO film	39	12	0.057
G 10*	Copper only	39	12	0.057
G 11	Copper + Cu_2O/CuO film			30
G 12	Copper + Cu_2O/CuO film		18	

* Group 9 and Group 10 were immersed in tap water, concentrations of anions were determined in our laboratory.

All reagents were of analytical (AR) grade unless otherwise stated. Solution and sample preparations, and washing of glassware were made with Milli-Q water. All experiments were performed at room temperature.

Potentiostatic Method

Potentiostatic experiments were performed on a potentiostat/galvanostat designed and built within the Faculty of Science and Technology at the University of Western Sydney, Nepean. The electrochemical cell consisted of the copper sheet with or without oxide layer as the working electrode, a platinum sheet as the auxillary and the reference was a saturated calomel electrode with a double junction. To minimise solution contamination the junction was filled with 2 M KNO_3 solution.

The preparation of the copper specimens were as described in another study [21], except that the surface modification was performed by exposure to 1.0 M sodium hydroxide solution for four and half days. Only 3 cm length of the specimen was exposed to the test solution and the total exposure area for both sides was 15 cm^2. Optical microscopic examination was performed at the end of the potentiostatic experiments to examine the effect of the potentiostatic polarisation on the copper surfaces.

Surface Analysis

A Rigaku D/max-1B (1.5 KW) X-ray Diffractometer was used to identify and characterize the chemically modified copper surfaces under control chemical conditions. Analytical conditions include a copper target, Cu $K\alpha_1$ wavelength of 1.540 Å with nickel filter, tube voltage and current of 40 kv and 30 mA, respectively. An International Scientific Instruments scanning electron microscope, 100A with PGT system 4 X-ray analyser utilising a Si(Li) detector was used for the morphological study of the modified copper surfaces.

3 RESULTS AND DISCUSSION

Electrochemical and Chemical Modification of Copper Surfaces

The cyclic voltammogram shown in Figure 1 indicates that the electrochemical formation of copper (I) and copper (II) oxides can be readily accomplished in basic media, as has been previously established [4,15]. The selective or preferential formation of cuprous oxide occurs at pH $\geq$ 4.8, whereas cupric oxide is formed at pH $\geq$ 6.7. The identification of both products on electrochemically polarised surfaces has been successfully accomplished by the use of X-ray diffraction analysis [15]. Table 2 gives the typical film composition on electrochemically polarised copper surfaces. The influence of solution pH and temperature on the film composition is clearly evident from these results. In general, the tendency for the formation of copper (I) oxide increased as the pH was lowered to about 5,

Table 2 The film composition on electrochemically modified copper surfaces in various aqueous media

pH	Temperature (°C)	
	25	100
4.95	Cu_2O	Cu_2O, CuO
4.95*	Cu_2O, CuCl	Cu_2O, CuO
7.03	Cu_2O, $6CuO{\cdot}Cu_2O$	Cu_2O, CuO $6Cu0.Cu_20$
11.70	-	Cu_2O, CuO
12.00	Cu_2O, CuO	-

* Formed in presence of Cl^- ions.

whereas the formation of copper (II) oxide predominated at the higher pH range. However, the need for a potentiostatic control of the surface to achieve adequate polarisation can represent a major hinderance in promoting the practical utilisation of the electrochemical surface modification approach. Ideally, the development of a more simplified approach for the formation of the oxides will increase the practical application of the approach. The utilisation of a chemical surface modification approach based on the immersion of copper surfaces in sodium hydroxide solution (1.0 M), resulted in the selective formation of copper (I) oxide within the first three days and gradually led to the simu-

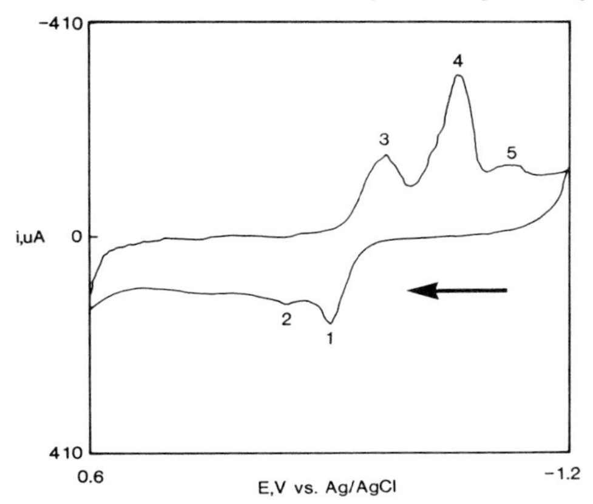

Figure 1 Cyclic voltammogram obtained for copper in 0.1 M sodium hydroxide solution, room temperature, 50 mV/s

ltaneous formation of copper (II) and copper(I) oxides upon further exposure to this medium. The substantial morphological changes which occurred on the copper surfaces in this medium over the ten-day exposure period was investigated by scanning electron microscopy. In the first four days, the copper surface was modified by a tightly packed thin film (reddish-brown product) which adhered strongly to the surface, but from the fifth day of exposure the formation of an additional top layer of a filamentous oxide became evident. The surface layer became more uniform after 2 days of exposure and the crystal structure was generally identical, indicating a fairly even distribution of the copper (I) oxide on the metal surface. From the fifth day, the relative proportion of the filamentous product increased with further exposure, and the colour of the oxide layer changed to greyish-black, covering most of the surface after 9 days of exposure. The X-ray diffraction analysis of the various oxides on the copper surfaces over the ten-day period revealed that copper (I) oxide was the dominant product in the first four days, but the presence of copper (II) oxide on the copper surface became evident from the fifth day. This latter oxide became the dominant product towards the end of the exposure period. It is interesting to note that there are obvious differences in the electrochemical and chemical modification approaches employed in this solution. For example, while the electrochemical polarisation of copper surfaces resulted in an instantaneous formation of copper (I) and copper (II) oxides, the chemical modification was a much slower process, enabling the selective or simultaneous formation of relatively more stable oxides.

Corrosion Resistance of Modified Copper Surfaces

Figure 2 illustrates that the presence of a mixed copper (I) and copper (II) oxide layer reduced the extent of the corrosion of copper in tap water considerably. In order to identify the factors responsible for the higher weight changes obtained for the bare copper specimen the influence of anions commonly present in tap water such as bicarbonate, carbonate and chloride ions was investigated.

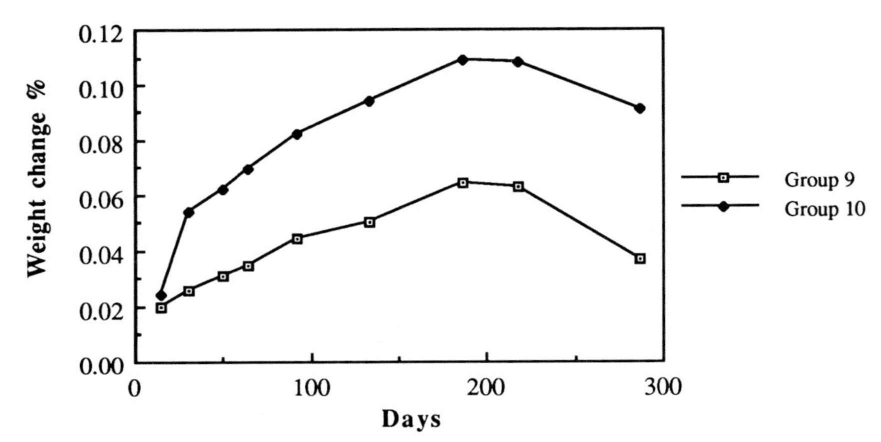

Figure 2 Influence of chemical surface modification on the long-term corrosion of copper in tap water (Water composition and surface conditions are as given in Table 1)

Figure 3 illustrates the influence of varying chloride concentration on the weight change of the modified copper surfaces. As clearly evident, the increasing chloride concentration resulted in a considerable weight change. It is interesting to note, from Figure 4, that the presence of higher bicarbonate concentration reduced the extent of the weight change. This observation indicates that bicarbonate ion concentration is critical in improving the corrosion resistance of the modified copper surfaces. The critical bicarbonate concentration for maintaining little or no weight change was 61 mg/l. For example, the approximate doubling of the bicarbonate concentration from 31 mg/l to 61 mg/l, as illustrated in Figure 4, resulted in a considerable reduction of the weight change. Similarly, an increase in the bicarbonate concentration to 305 mg/l reduced the weight change considerably.

In addition to the little or no weight change observed for specimens 5 and 8, the morphological study of the surfaces revealed that the oxide layer remained stable. Visual observations indicate that specimen 5 was more stable. The X-ray diffraction results in Table 3 indicate that the number of the surface components increased with an increase in the exposure period. In most cases where bicarbonate and/or carbonate ions are present, the formation of basic copper carbonate occurred. It was interesting to note that in two cases (specimens 5 and 8), where the bicarbonate ion concentrations were relatively higher, the basic copper carbonate was not present. In effect the original oxide components remained on the modified copper surfaces throughout the long-term

Table 3 X-ray diffraction analysis of the chemically modified surfaces after exposure to the different anions

Group	55 Days	212 Days	287 Days
G 1	Cu_2O (3), CuO (3)	Cu_2O (3), CuO (3)	Cu_2O (3), CuO (3) , $CuCO_3Cu(OH)_2$ (3)
G 2		Cu_2O (3), CuO (3) , $CuCO_3Cu(OH)_2$ (3)	Cu_2O(3), CuO(3) $CuCO_3Cu(OH)_2$ (3)
G 3	Cu_2O (3)	Cu_2O (3), CuO (1)	Cu_2O (3), CuO (2), $CuCO_3Cu(OH)_2$ (3)
G 4	Cu_2O (3), CuO (1)	Cu_2O (3), CuO (1) , $CuCO_3Cu(OH)_2$ (3)	Cu_2O (3), $CuCO_3Cu(OH)_2$ (3)
G 5	Cu_2O (3), CuO (1)	Cu_2O (3), CuO (1)	Cu_2O (3), CuO (1)
G 6	Cu_2O (3)	Cu_2O (3), CuO (2)	Cu_2O (3)
G 7	Cu_2O (3), CuO (3) , $CuCO_3Cu(OH)_2$ (3)	Cu_2O(3), CuO (3) $CuCO_3Cu(OH)_2$ (3)	Cu_2O (3), CuO (3) $CuCO_3Cu(OH)_2$ (3)
G 8	Cu_2O (3), CuO (1)	Cu_2O (3), CuO (1)	Cu_2O (3), CuO (1)
G 9	Cu_2O (3), CuO (3)	Cu_2O (3), CuO (3)	Cu_2O (3), CuO (1)
G 10		Cu_2O (3)	Cu_2O (3), CuO (1)
G 11	Cu_2O (3), CuO (3)	Cu_2O (3), CuO (2) $CuCO_3Cu(OH)_2$ (3)	Cu_2O (2), CuO (3) $CuCO_3Cu(OH)_2$ (3)
G 12	Cu_2O (3), CuO (1)	Cu_2O (3), CuO (1)	Cu_2O (3)

*(No.) indicates the number of most intense diffraction lines identified.
#The modified surfaces gave Cu_2O (3) and CuO (1) prior to immersion.

exposure period. Thus, indicating that the presence of > 61 mg/l bicarbonate improves the stability of the copper (I) and copper (II) oxides on the chemically modified copper surfaces.

Potentiostatic Polarisation Measurement

The potentiostatic polarization measurement for the modified copper specimens in 0.005 M $NaHCO_3$ solution demonstrated a considerable difference in the steady-state current obtained at different applied potentials. The application of 842 mv vs SCE enabled the attainment and retention of low steady-state current. Optical microscopic examination of the modified copper surface after polarisation in the

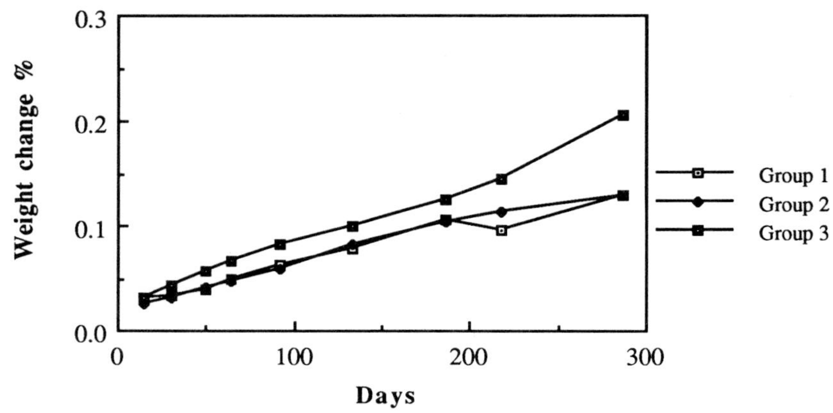

Figure 3 Influence of increasing chloride concentration on the corrosion resistance of the chemically modified copper surfaces

higher sodium bicarbonate solution (0.05 M) revealed that the original surface condition was maintained. X-ray diffraction analysis of the surface also revealed that only copper (I) and copper (II) oxides were present. In contrast, a much higher steady-state current density and loose green/blue products were obtained on the surface of a bare copper specimen during a similar potentiostatic measurement. The corrosion product responsible for the green/blue substance was identified by XRD as $Cu(OH)_2CO_3$.

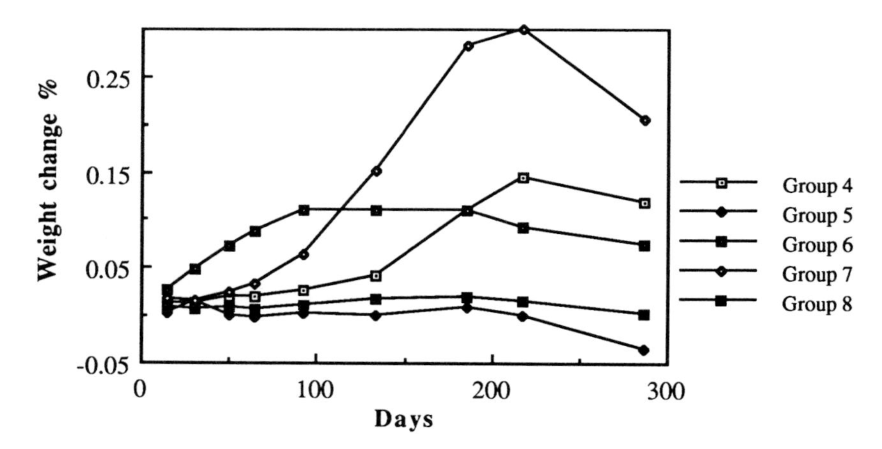

Figure 4 Influence of bicarbonate concentration on the corrosion resistance of modified copper surfaces

Figure 5 indicates that the addition of chloride resulted in a considerable instability of the oxide layer in the 0.05 M sodium bicarbonate solution. The rapid increase in the stable-state current density in this case appeared to be due to the initiation of the formation of the basic copper carbonate by the presence of chloride ions. The use of lower bicarbonate concentration (0.0005 M) also resulted in the increasing current density. Thus indicating that the stability of the chemically modified surface is reduced in the dilute solution.

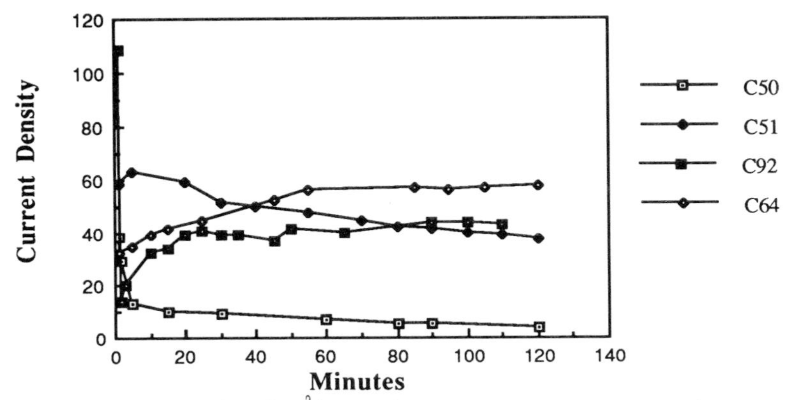

Figure 5 Current ($\mu A/cm^2$) - time measurement on the modified surfaces in sodium bicarbonate solutions: C50 0.05 M, C51 0.005 M, C92 0.05 M + 0.01 M NaCl, C64 0.0005 M, applied potential 842 mV vs SCE

4 CONCLUSION

The chemical modification of copper surfaces in sodium hydroxide solutions has been found to be useful in improving the corrosion resistance of the metal in tap water. The presence of bicarbonate ions improved the stability of the the oxide layer considerably and resulted in improved corrosion resistance. In contrast, the presence of chloride reduced the stability of the oxide layer by initiating the formation of basic copper carbonate on the metal surface. Evidently, the reduction of the chloride effect on the modified copper surfaces requires the addition of bicarbonate ions.

ACKNOWLEDGEMENT

The authors are grateful to Curt Stocksiek for his assistance with the XRD measurements, Alex Hons for his assistance with SEM measurements, and to the University for the provision of a Research Seed Grant and a Summer Vacation Research Scholarship for this project.

REFERENCES

1. F.J. Cornwell, G.Wildsmith and P.T. Gilbert, Br. Corros. J., 1973, 8, 202.

2. M. Akkaya and J.R. Ambrose, Mat. Perf., 1987, 26, (3), 9.
3. E. Mattsson, Mat. Perf., 1987, 26, (4), 9.
4. S.B. Adeloju and H.C. Hughes, Corros. Sc., 1986, 26, (10), 851.
5. M. Pourbaix, 'Lectures on Electrochemical Corrosion', Plenum Press, New York, 1973.
6. M. Linder, KI Report, (2), 1987.
7. T. Fujii, T. Kodama, et al., Corros. Sc., 1984, 24, (10), 901.
8. A. Rodolfo, Jr. et al., J. Am. Wat. Works Assoc., 1987, 79, (2), 62.
9. A. Cohen and J.R. Myers, J. Am. Wat. Works Assoc., 1987, 79, (2), 58.
10. H. Steve, J.F. Reiber and M.M. Menjamin, J. Am. Wat. Works Assoc., 1987, 79, (2), 71.
11. C. Deslouis and B. Tribollet, J. Appl. Electrochem., 1988, 18, 374.
12. J.L. Ord, D.J. Desmet and Z.Q. Huang, J. Electrochem. Soc., 1987, 134, (4), 826.
13. M.R. Gennero De Chialvo, J.O. Zerbino and S.L. Mauchiano, J. Appl. Electrochem., 1986, 16, 517.
14. L.D. Burk, M.J. Ahern and T.G. Ryan, J. Electrochem. Soc., 1990, 137, (2), 553.
15. S.B. Adeloju, Mikrochimica Acta, 1986, III, 401.
16. K. Nassau, P.K. Gallagher, A.E. Miller and T.E. Graedel, Corros. Sc., 1987, 27, (7), 669.
17. H.H. Strehblow and B. Titze, Electrochim. Acta, 1980, 25, 839.
18. D.W. Shoesmith, T.E. Rummery, D. Owen, and W. Lee., J. Electrochem. Soc., 1976, 123, (6), 790.
19. S.M. Wilhelm, Y. Tanizawa, C.Y. Liu, N. Hackerman, Corros. Sc., 1982, 22, (8), 791.
20. C.H. Pyun, S.M. Park, J. Electrochem. Soc., 1986, 133, (10), 2024.
21. S.B. Adeloju and Y.Y.Duan, Br. Corros. J., Submitted for publication, (1992).

2.2.6
Inhibition of Copper Corrosion in Tap Water with Sodium Propionate

S.B. Adeloju and T.M. Young

ELECTROCHEMICAL AND CORROSION TECHNOLOGY RESEARCH GROUP, DEPARTMENT OF CHEMISTRY, UNIVERSITY OF WESTERN SYDNEY, NEPEAN, PO BOX 10, KINGSWOOD, NEW SOUTH WALES 2747, AUSTRALIA

1 INTRODUCTION

Copper has gained a lot of uses over the years because of its excellent resistance to corrosion, high electrical and thermal conductivity, malleability and availability. The accomplishment of substantial protection with this metal is attributed to the formation of sparingly soluble, adherent and impervious corrosion products on the surface in most aqueous media. Although failures are not common, this may result from pitting and erosion corrosion[1-3], and as such it is often necessary to implement some protection strategies under these circumstances.

Effective protection against the corrosion of copper may be provided by the use of an inhibitor which enables a significant reduction of the corrosion processes. Currently, one of the most widely used inhibitor for the control of copper corrosion in various aqueous media is benzotriazole[4-6]. This inhibitor is effective and efficient at small concentrations of about 1% w/v. However, there is still a need for the development or identification of other non-toxic substances which are more effective either at lower concentrations or which are relatively less expensive.

In recent years one of the emerging group of substances which has demonstrated some potential for the inhibition of metallic corrosion in aqueous media are carboxylates[7-14]. To date, only limited studies have been reported on the inhibition of copper corrosion by use of these substances. One relevant study in this area was carried out over four decades ago by Wormwell and Mercer[7]. This study suggested that effective inhibition of copper corrosion in aqueous media could be accomplished with as low as 0.05% sodium benzoate. Other related studies include the inhibition of steel corrosion with propionic acid and the significance of organic chain length[8], protective film formation[9,10], the mechanism of retardation[10,11], reduction of the pitting corrosion of iron with low concentration of sodium propionate[12], and the use of acetate, cinnamate and nitrocinnamate for inhibition of steel corrosion[13-16].

In this study, the effectiveness of sodium propionate for the inhibition of copper corrosion in tap water will be investigated. This will involve the use of weight-loss, potential-time, polarisation resistance and potentiodynamic polarisation methods for the evaluation of the effectiveness and mechanism of protection afforded by the inhibitor. In addition, X-ray diffraction analysis will be employed for the identification of the protective film on copper surfaces. Furthermore, the influence of pH on the effectiveness of the inhibitor will be examined.

2 EXPERIMENTAL PROCEDURE

Electrochemical Measurement

A three electrode cell consisting of a copper working electrode, an Ag/AgCl reference electrode and a platinum sheet auxillary electrode was used in conjuction with a UTAH potentiostat for all electrochemical measurements. The end of the luggin capillary which housed the Ag/AgCl reference was filled with 3% agarose gel made up in NH_4NO_3. This gel has been proven to be stable, even at high temperatures[17]. The working electrode was prepared by fitting a cylindrical copper specimen into cold curing acrylic resin. The end of the rod (0.1 cm^2 in area) was used as the working surface of the electrode. Deaeration of the test solution was accomplished by bubbling nitrogen into the test solution for 30 minutes prior to conducting the electrochemical measurements. A nitrogen atmosphere was also maintained over the solution during the measurement. Prior to immersion in the test solution, the electrode surface was polished with abrasive paper grades of 360 and 1000, rinsed with tap water, degreased with acetone, and finally rinsed with the test solution.

Weight Loss Measurement

Coupon Preparation. Copper coupons (99.9% purity) of 30 cm^2 in area were prepared for weight loss measurements by the ASTM standard method (G1-81) for preparation and cleaning specimens. The samples were abraded using grades P100, P360 and P1000 abrasive paper, degreased with acetone, dried with a clean tissue and placed in a dessicator until required for use. The sample was weighed prior to being suspended in the test solution.

Immersion Test. A 9.6% w/v stock solution of the inhibitor was prepared by dissolving 9.6000 grams of sodium propionate in 30 cm^3 of methanol and made to 100 cm^3 with Milli-Q water. Weight loss measurements of coupons were made in two litres of test solution containing tap water and inhibitor at thirty, sixty and ninety days intervals. The tap water in each beaker was topped up regularly to compensate for losses due to evaporation. These experiments were performed under identical conditions with or without stirring. At the end of the exposure period, the coupons were taken out, washed, and finally reweighed. The difference between this and the initial weight was used to calculate the weight-loss. Where necessary, the pH of the solution was adjusted to 1, 5, 10 and 12 using either concentrated sodium hydroxide or nitric acid. A 20-30 cm^3 of the solution is required for 1 cm^2 of coupon [14].

X-ray Diffraction Analysis

The films formed by the exposure of copper coupons to deionised water containing 0.10% w/v sodium propionate at pH 5 and 12 were analysed by XRD using the following conditions: 40kV, 30mA; copper target at a wavelength of 1.5406 A°, divergence slit of 1°, recieving slit of 0.15°, scan speed of 1 degree/min and sampling at 0.02°. The sample was mounted into the diffractometer (Rigaku /Geigerflex D/Max, 15kW Series). The resulting diffractograms were interpreted by the use of the search match program and search manuals (Powder Diffraction Files for Inorganic and Organic Compounds, JCPDS 1988).

3 RESULTS AND DISCUSSION

Influence of pH

The extent of corrosion of the copper coupons in deionised water varied considerably with pH. As can be expected, the coupons dissolved completely within four days of exposure to a solution of pH 1. In contrast, the corrosion rates at pH $\geqslant 5$ decreased progressively over the ninety-day exposure period, possibly due to the formation of a passive layer. The best corrosion resistance in the deionised water was accomplished at pH 5. In the presence of 0.10% w/v sodium propionate the corrosion rates decreased gradually in solutions within pH 1-12 over the ninety-day period. The data in Table 1 indicate that the optimum pH range for the effective inhibition of copper corrosion by sodium propionate in the deionised water lies between 7 and 12. The weight gain obtained at pH 12 may be accounted for by the formation of a protective oxide, as well as the adsorption of the inhibitor. These observations suggest that the presence of cupric oxide, identified by XRD, on the surface of the coupon is more favourable for effective inhibition of copper corrosion than the cuprous oxide formed at pH 5. The different behaviour of the inhibitor on these surfaces may be associated with the different morphology of the oxide layers on the copper surfaces. The cuprous oxide layer, being a tightly packed and more strongly adhered film, may enable only limited interaction of the inhibitor with the metal, whereas the cupric oxide which appeared filamentous and porous in nature may enable more interaction with the substance.

Weight Loss in Tap Water

The surfaces of copper coupons immersed in tap water without inhibitor and in those containing 0.025% and 0.050% w/v sodium propionate had slightly black-dull finish, whereas no apparent attack was evident on those in tap water containing 0.10% w/v of the inhibitor. However, the weight loss measurements over the 90-day period did not result in any significant weight gain or weight loss. This is clearly indicative of the good corrosion resistance of copper in the tap water. The visual observations of the copper surfaces indicate that the mechanism of corrosion inhibition is very much dependent on the inhibitor concentration. In the presence of sodium propionate concentrations $< 0.10\%$ w/v, the inhibition was due partly to the formation of black oxide on the copper surface, whereas at 0.10% w/v the retention of the shiny copper surface suggests that the adsorption of the inhibitor may play a greater role in this case. The observation is particularly significant as the avoidance of the oxide formation by the inhibitor may be useful in reducing or eliminating the incidence of pitting corrosion, which is often associated with the breakdown of surface oxide layer.

Table 1 Corrosion rates of copper in deionised water containing 0.10% w/v sodium propionate at different pH values.

Solution pH	Exposure Period		
	Corrosion Rate for 30 Days ($g/m^2/day$)	Corrosion Rate for 60 Days ($g/m^2/day$)	Corrosion Rate for 90 Days ($g/m^2/day$)
1	1.9330	3.1920	2.6430
5	0.7150	0.3980	0.2455
7	0.0180	0.0100	0.0069
10	0.0090	0.0050	0.0043
12	-0.0190	-0.0050	-3.0200

The results in Table 2 reveal that as little as 0.025% w/v sodium propionate was effective in reducing copper corrosion under stirring condition. Again the best inhibition was accomplished in the presence of 0.10% w/v of the inhibitor.

Potentiodynamic Polarisation

Figure 1a shows the anodic polarisation curve for a naturally-aerated tap water. The addition of 0.025% w/v of the inhibitor had little effect on the anodic process (Figure 1b). However when the amount of inhibitor was increased to 0.050% and 0.10% w/v sodium propionate the anodic process was altered, as shown in Figures 1c and 1d . The current was reduced substantially as the concentration of the inhibitor was increased and the passive regions are evident. At a critical potential beyond the passive region, a sharp increase in current was evident. Based on the nature of the corrosion system it is appropriate to refer to this potential as the 'desorption potential, E_{des}', instead of the breakdown potential as it is normally referred to in the case of pitting corrosion of steel in solutions which contain activating ions such as halogens. The desorption potential, E_{des}, remained consistently at approximately 200 mV vs Ag/AgCl, but with slight increase in the positive direction as the inhibitor concentration was increased.

The observed shift in the potential towards the positive direction when oxygen is present is due to its cathodic reduction and this is more favourable for the adsorption of the inhibitor and/or maintenance of a

Table 2 Corrosion rates of copper coupons immersed in stirred tap water with and without sodium propionate for ninety days.

Inhibitor Concentraion	Average Weight Loss (mg)	Corrosion Rate ($g/m^2/day$)
Tap water	2.200	0.0095
+ 0.005%	2.100	0.0091
+ 0.010%	3.830	0.0165
+ 0.025%	1.300	0.0056
+ 0.050%	1.300	0.0056
+ 0.10%	1.000	0.0043

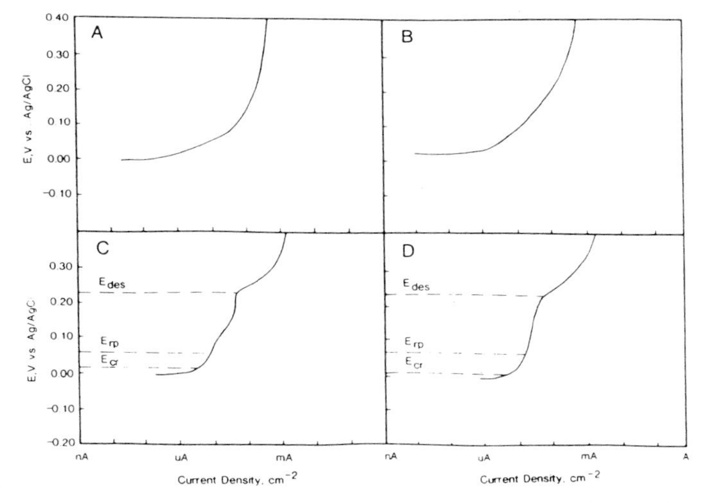

Figure 1 Anodic potentiodynamic polarisation of copper in (a) naturally aerated tap water, (b) +0.025%, (c) +0.05%, and (d) +0.10% w/v inhibitor

protective oxide. Evidently, a shift of the potential to the negative region in the deaerated solution was due to the absence of oxygen and, hence, to the elimination of the cathodic process. Due to the depletion of oxygen in this system it was necessary to apply some external polarisation to shift the potential to a region where passivation is possible. Figure 2 shows typical potentiodynamic polarisation curves obtained for copper in tap water in the absence of inhibitor and in those containing different concentrations of sodium propionate after deaeration, respectively. It is evident from these results that copper demonstrates some passivation in the tap water, even in the absence of the inhibitor. This observation is consistent with the previous suggestion from the weight-loss measurements. However, the extent of passivation was much more pronounced in the presence of the inhibitor at > 0.025% w/v.

In general the current increased with an increase in the potential until a diffusion limited regime is reached. Evidently, copper is active in the deaerated tap water and similar behaviour was observed with the addition of 0.025% w/v sodium propionate. However, with the addition of 0.050% and 0.10% w/v sodium propionate a reduction in current with an increase in concentration of the inhibitor was observed. Again, a breakdown in passivity occurred at approximately 200 mV vs Ag/AgCl, and hence an increase in current is apparent. In contrast to the naturally aerated solutions, the desorption potential (E_{des}) decreased slightly with an increase in the inhibitor concentration. The increase in the inhibitor concentration also shifted the critical potential to more negative value and, hence, indicate better ease of passivation.

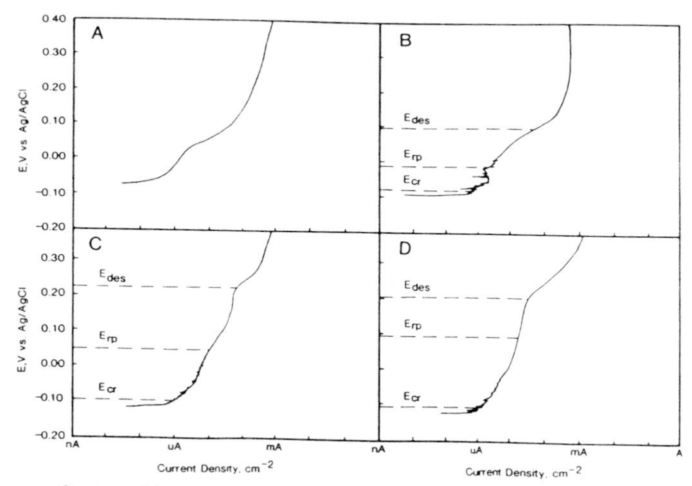

Figure 2 Anodic potentiodynamic polarisation of copper in (a) deaerated tap water, (b) +0.025%, (c) +0.05%, and (d) +0.10% w/v inhibitor

Corrosion Potential-Time Measurements

Figure 3 shows the potential-time curves for copper in tap water containing different concentrations of sodium propionate. In all cases, the potential of copper remained in the positive region after exposure for 400 hours. The extent of passivation increased with increasing inhibitor concentration and exposition time, as demonstrated by the increasing positive potential. The most positive potential and, hence, the best passivation was obtained in the presence of 0.10% w/v sodium propionate. The shift and maintenance of the copper potential in the positive region confirms earlier suggestions that passivation occurs in the tap water, even in the absence of the inhibitor. However, the extent of passivation was more pronounced in the presence of the inhibitor .

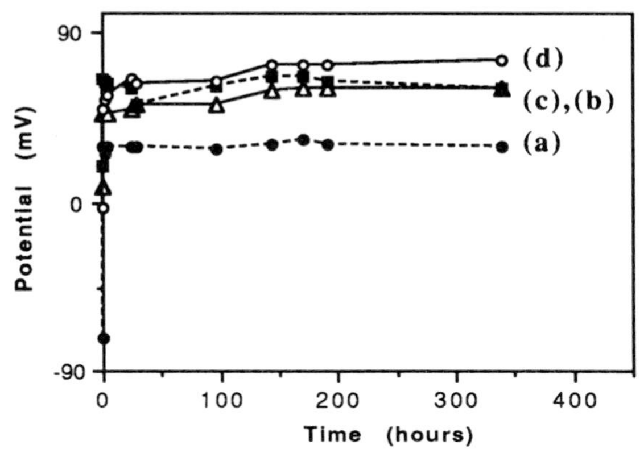

Figure 3 The dependence of copper corrosion potential on time in (a) tap water only and tap water containing (b) +0.025%, (c) +0.05%, and (d) +0.10% inhibitor

Polarisation Resistance Measurements

The effect of the inhibitor on the polarisation resistance of copper in naturally aerated tap water is shown in Figure 4. In the absence of the inhibitor the polarisation resistance remained constant over the three hour period due to the presence of a stable oxide on the copper surface,

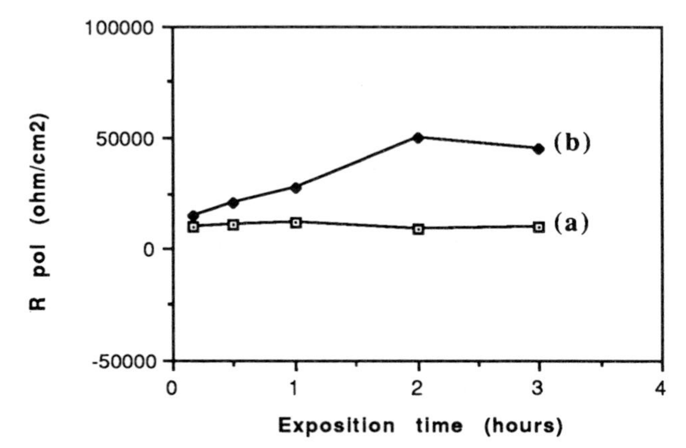

Figure 4 Polarisation resistance of copper in (a) naturally aerated tap water and (b) +0.10% w/v inhibitor

whereas the addition of the inhibitor resulted in an increase in the polarisation resistance with time. Evidently, the increased polarisation resistance was due to passive film formation which resulted from either the increasing adsorption of the inhibitor and/or copper oxide formation. The stabilisation of the polarisation resistance of copper in the presence of the inhibitor towards the end of the measurements indicates that maximum passivation is accomplished after two hours.

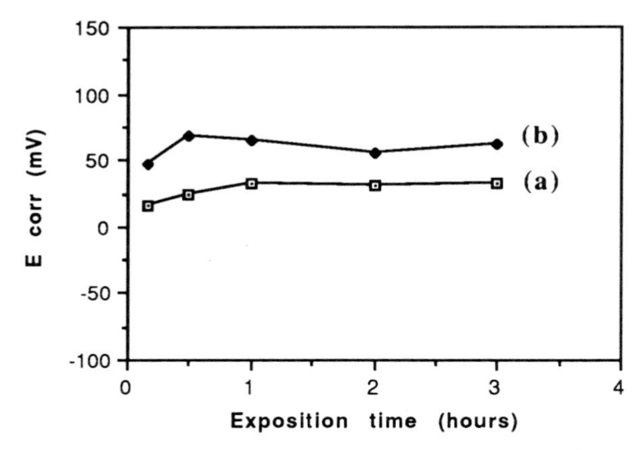

Figure 5 Influence of inhibitor on the corrosion potential of copper in (a) naturally aerated tap water without inhibitor and (b) with 0.10% w/v inhibitor

Figure 5 shows that the corrosion potential obtained during the measurement of the polarisation resistance shifted in the positive direction in the presence and absence of the inhibitor. This observation also confirms the earlier suggestion that a passive film is formed on the copper surface in the tap water, even in the absence of the inhibitor. As

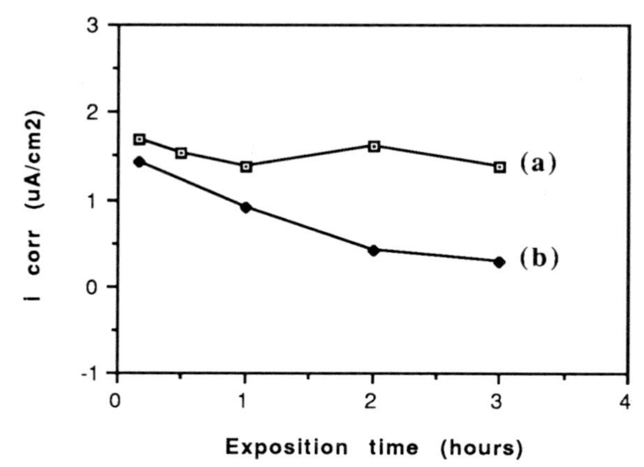

Figure 6 Influence of inhibitor on the corrosion current of copper in (a) naturally aerated tap water without inhibitor and (b) with 0.10% w/v inhibitor

demonstrated by the more positive potential obtained in the presence of the inhibitor, it is clearly evident that the corrosion resistance of copper is further improved under this condition.

Figure 6 also shows that the corrosion current of copper in tap water was substantially reduced with the addition of the inhibitor. The corrosion currents obtained in the tap water approached a steady-state value after 30 minutes, whereas in the presence of the inhibitor a slow decline in current was evident, reaching a steady-state value after 2 hours. The attainment of the steady-state corrosion current over this period is consistent with the observation made from the polarisation resistance and corrosion potential measurements, and further supports the view that maximum passivation is accomplished after 2 hours exposition to tap water containing 0.10% w/v inhibitor.

X-ray Diffraction Analysis

The diffractograms obtained for the passive layer on copper surfaces in deionised water containing 0.10% w/v of the inhibitor at pH 5 and 12 indicate, as demonstrated by the data in Table 3, that cuprous oxide, Cu_2O, was formed at pH 5, while at pH 12 cupric oxide, CuO, was the only identified product. The intensity and the positions of the three most intense lines are in reasonably good agreement with those quoted for the pure compounds in Table 3. The incorporation of the inhibitor was evident from the greenish-black coloration of the cuprous oxide.

Table 3 XRD data for corrosion products of copper in deionised water containing 0.10% w/v inhibitor at pH 5 and 12.

COMPOUND NAME	COLOUR	STRONGEST DIFFRACTION LINES (JCPDS)	EXPERIMENTAL LINES
pH 5 CUPROUS OXIDE [Cu_2O]	RED BROWN	2.465 (100)	2.444 (15)
		2.135 (37)	2.120 (3)
		1.510 (27)	1.504 (4)
pH 12 CUPRIC OXIDE [CuO]	BLACK	2.523 (100)	2.516 (2)
		2.323 (96)	2.322 (2)
		1.806 (25)	1.805 (29)

4 CONCLUSION

The use of sodium propionate for the inhibition of copper corrosion in aqueous media has been successfully demonstrated. The formation of a cupric oxide under-layer at pH 12 was more favourable than that of cuprous oxide formed at pH 5 for the inhibition of copper corrosion in tap water. The required inhibitor concentrations for the effective prevention of copper corrosion under stirring and stagnant conditions in the tap water are 0.025 and 0.10 % w/v, respectively.

ACKNOWLEDGEMENT

The authors are grateful to Curt Stocksiek for his assistance with the XRD measurements, and to the University for the provision of a Research Seed Grant and a Summer Vacation Research Scholarship for this project.

REFERENCES

1. A. Cohen and J.R. Myers, J. Am. Wat. Works Assoc., 1987, 2, 58.
2. F.M. Al-Kharafi, H.M. Shalaby and V.K. Gouda, Br. Corros. J., 1989, 24, (4), 284.
3. H.M. Shalaby, F.M. Al-Kharafi and V.K. Gouda, Corrosion, 1989, 45, (7), 536.
4. R. Cigna, K. De Ranter, et al., Br. Corros. J, 1988, 23, (3), 190.
5. S.F. Da Costa, S.M. Agostinho and J.C. Rubim, J. Electroanal. Chem., 1990, 295, (1-2), 203.
6. H.C. Shih and R.J. Tzou, J. Electrochem. Soc., 1991, 138, (4), 958.
7. F. Wormwell and A.D Mercer, J. Applied Chem.(London) , 1952, 2,150.
8. E. Constantinescu and E. Heitz, ' Corrosion Science ', 1976, 16, (11), 857.
9. M. Stern, J.Electrochem. Soc. , 1958, 105, (11),638.
10. P. Spinelli, R. Fratesi and G. Roverti, Werkstoffe und Korrosion, 1983, 34, 161.
11. R.D Granata, P.C Santiesteban and H. Leidheiser Jr, Proc. Electrochem. Soc. (Surface Inhibition and Passivation), 1986-7, 69.
12. S.M. Abd El Haleem, M.G.A. Knedr and A.A Abdel Fattah, Proc. Int. Cong. Metal Corr., 1984, 4, 362.
13. G. Pallos-Leitner, 'Evaluation of Corrosion Inhibitors With Particular Reference to Australian Water', PhD Thesis, University of New South Wales, 1982.
14. I.L Rosenfeld, 'Corrosion Inhibitors', translated by Ron and Hilary Hardin, McGraw Hill, New York, 1981.
15. R.D. Granata, P.C. Santiesteban and H. Leidheiser,Jr., Proc. 6th Euro.Symp. on Corr. Inhib., 1985, 1, 411.
16. M.S. Aal and M.H. Wahdan, Br. Corros. J., 1981, 16, (4), 205.
17. S .B. Adeloju, Mikrochimica Acta, 1986, III, 401.

2.2.7
An Assessment of the Corrosion Protection Offered to Steel and Aluminium Alloys by Aluminium-based Metal Sprayed Coatings When Exposed to South African Mine Waters

S.M. Ford, F.P.A. Robinson, and J.E. Leitch[1]

DEPARTMENT OF METALLURGY AND MATERIALS ENGINEERING, UNIVERSITY OF THE WITWATERSRAND, JOHANNESBURG, SOUTH AFRICA

[1] HULETT ALUMINIUM (PTY) LTD., PIETERMARITZBURG, SOUTH AFRICA

1 INTRODUCTION

Aluminium alloys are used extensively in the construction of shaft conveyances for transporting miners to working levels in South African gold mines. The shafts are often wet with fissure waters, some of which are fairly corrosive towards carbon steel and aluminium. This corrosivity is exacerbated by the hybrid construction of the cages, which utilize carbon steel for fasteners and critical structural parts together with aluminium alloy profiles and plates. This results in the creation of numerous dissimilar metal couples.

Conventional methods for the elimination of galvanic corrosion, such as insulating membranes or paint coatings, have had varied success, due to diverse factors, such as destruction of membranes by vibration or erosion caused by falling rocks. In aggressive waters, the aluminium plates which are typically AA5083 alloy, can be subject to galvanic corrosion in areas, for example, around steel bolt heads and swaged collars. This corrosion manifests itself as localized pits, blistering or exfoliation, which are characteristic of this 5% magnesium alloy.

The purpose of this work was to develop sprayed aluminium coatings which are suitably anodic to the required aluminium substrate alloys so as to protect sensitive components. These coatings should also protect carbon steel, and this aspect was thus also evaluated.

The Fundamentals of Anodic Metal Spray Coatings

The purpose of an anodic metal sprayed layer is not merely to isolate the substrate from the environment, but also to act as a sacrificial anode at regions where the substrate is exposed.[1] Figure 1 below illustrates the action of an anodic coating in sacrificially protecting the substrate at a discontinuity, with consequent consumption of the coating. If this progresses too far and too fast, the coating is no longer able to act as an effective barrier or to protect the substrate sacrificially at exposed areas some distance from the coating. The size of bare substrate that can be completely protected depends upon the "throwing power" of the coating, which is related to both its corrosion potential and the conductivity of the solution within the discontinuity.

Figure 1(g) demonstrates the situation that might prevail when the substrate is exposed at a discontinuity. Initially the coating corrodes, resulting in the formation of corrosion products in and around the area of bare substrate. As this process continues, so the corrosion products build up, creating a relatively impermeable barrier between substrate and environment, resulting in a reduced amount of sacrificial protection being required.

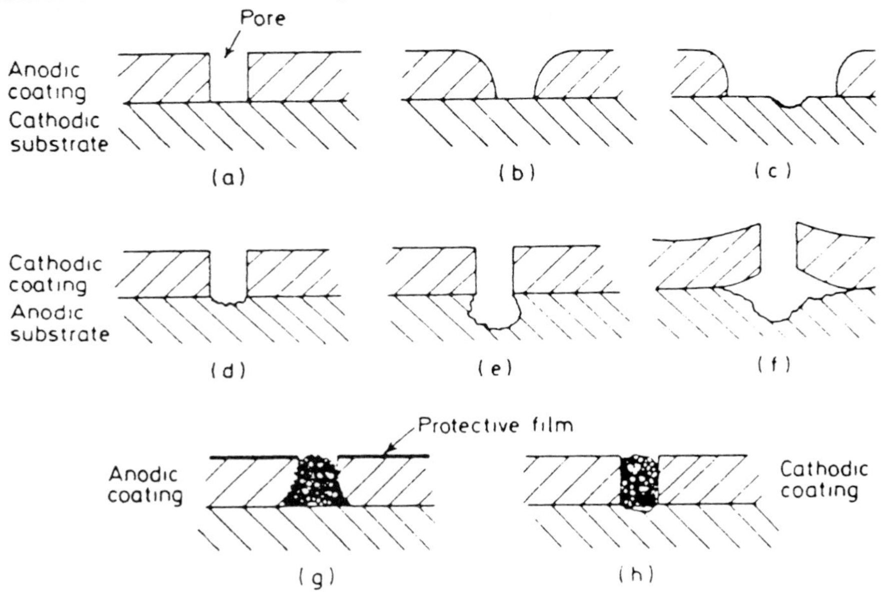

Figure 1 Diagrammatic representation of galvanic effects produced at a pore in coating.[1]

The Protective Action of Aluminium Alloy Coatings. From a comparison of the standard electrode potentials of aluminium and iron, the large potential difference suggests that pure aluminium would exert good sacrificial protection when coated onto iron-based alloys. In practice, however, this protection is somewhat stifled due to the stable oxide film on, and oxide content within, the metal sprayed coating which reduces the number of active sites on the surface.[1,2] When aluminium coatings are applied to aluminium alloy substrates a potential difference may still prevail due variations in alloy contents, but in general this potential is very small. It is possible however, to increase the anodic nature of the coating through additions of more active elements, such as zinc or magnesium to the aluminium coating.

Considering the merits of zinc-based coatings when compared to those of aluminium, the major advantage of the zinc is its higher chemical reactivity, thus enabling more effective cathodic protection to be afforded the substrate.[3] This advantage is countered by the associated higher corrosion rate and consequential reduced coating lifetime. Also, unlike aluminium, the corrosion products of zinc are more soluble and thus cannot effectively block the pores in the coating and provide the additional protection of aluminium.[4]

Previous investigations into the protection of aluminium alloy substrates using sprayed metal coatings are limited and on the whole very specific in nature.[5-7] Suggestions as to the coating compositions vary from author to author but when coating with aluminium-based alloys the major additions are usually zinc, magnesium or silicon. On the other hand, the metal spraying of iron-based substrates has a long commercial history and has been researched extensively.[8-13] Zinc is in this instance the most common alloying element with the percentage added to the aluminium either being very low (ie. 1-5%), or in excess of 50%.

<u>The Properties of Local Mine Waters</u>

The mine waters of South African gold mines are diverse in nature and often very corrosive due to their composition and/or the environment in which they are found. The aggressive chloride and sulphate ions frequently occur at levels exceeding 2500mg/l and others such as nitrates and ammonia are found in significant amounts. Analysis of a total of 46 samples of waters from eight different mines[14] showed that six samples had chloride contents above 2000mg/l, eight had sulphates above 1500mg/l, and five had pH below 5.0. Temperatures reach maxima of around 60°C and the waters are often contaminated with organisms, which give rise to microbial corrosion.

2 EXPERIMENTAL APPROACH

<u>Coating and Substrate Alloy Selections</u>

From a consideration of the effects of principal alloying elements on the electrolytic solution potential of aluminium alloys, it was decided that the following alloys would give a range of differences in potential relative to the alloy substrates.

(i) Pure aluminium
(ii) Aluminium/4.5% zinc
(iii) Aluminium/8% zinc
(iv) Aluminium/12% zinc

The four alloys were extruded and drawn down to 3.125mm wire, the required feedstock for the metal spray gun used.

Four materials were selected as substrate alloys, these being :

(i) AA6261 aluminium alloy
(ii) AA5083 aluminium alloy
(iii) A metal matrix composite (10% alumina in alloy 6061)
(iv) Mild steel.

Although the research was initially aimed at the coating of aluminium alloys, mild steel was included as all the fasteners on the mine cages would be made of iron-base alloys and would also be metal sprayed to prolong their operational life.

Test Coupon Preparation. Prior to spraying, all coupons were cleaned and then grit blasted with alumina. All sides of the coupons were sprayed, using an oxy-acetylene spray gun, and although no exact coating thickness could be achieved, a layer of roughly 100 microns of aluminium was measured on all the coupons.

In order to test the "throwing power" of the anodic coatings some coupons had defects of a controlled size drilled into their surfaces so that the coating was completely penetrated, leaving exposed substrate. Figure 2 demonstrates these contrived defects.

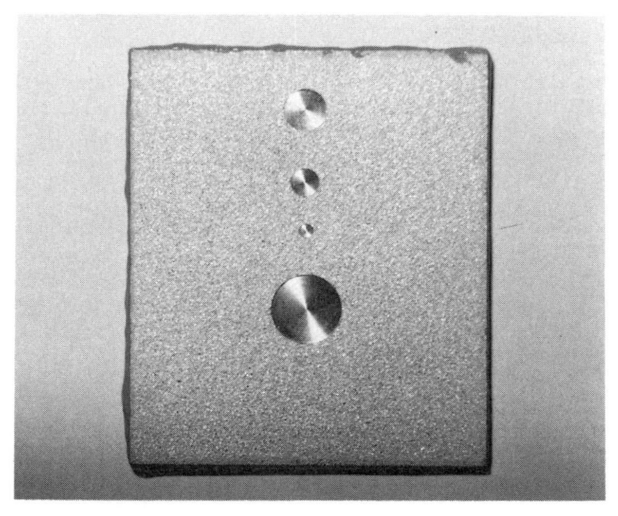

Figure 2 Controlled defects in test coupons

Test Solutions. Rather than using actual mine waters which would clog up corrosion testing apparatus on account of the high level of suspended solids, it was decided to make up synthetic waters. These were based on the chemical compositions of the more corrosive local mine waters. Three waters were selected with the compositions remaining the same, except for the sulphate and chloride levels which were varied to create the following combinations :

A. SO_4^{2-} - 2000mg/l
Cl^- - 2000mg/l

B. SO_4^{2-} - 500mg/l
Cl^- - 2000mg/l

C. SO_4^{2-} - 2000mg/l
Cl^- - 500mg/l

All the waters contained approximately 5000mg/l dissolved solids.

Furthermore, tests were conducted using a domestic potable water as a control, and 0.1N sulphuric acid to examine the behaviour of the various coating/substrate combinations in a highly corrosive environment.

Testing Techniques

Electrochemical Testing. Two accelerated electrochemical testing techniques were applied. Firstly, polarisation curves of the potential versus the current density were generated via potentiodynamic tests. Secondly, E_{corr} versus time tests established how the free corrosion potential varied with time. These tests were carried out in all three mine waters and in the municipal supply water.

Exposure Testing. The other approach taken to the testing was by means of various forms of immersion trials. Three basic modes of testing were used, namely :

(i) Salt spray (in accordance with ASTM B117 - but using a synthetic mine water)
(ii) Alternate immersion (in accordance with ASTM G44)
(iii) Total immersion (in accordance with ASTM G31).

3 RESULTS

Electrochemical Test Work

Potentiodynamic Tests. The relative performance of each coating is best related back directly to the actual potentiodynamic plot. Although each water gave rise to different plots, the trends established were the same. These trends are shown in Figure 3.

The superimposed scales clearly show the effect of the zinc additions on both the corrosion potentials and current densities of the coating alloys. Zinc, being more anodic than aluminium has the effect of lowering the corrosion potential and increasing the current density as higher percentages are added. This means that the more zinc in the coating, the greater the potential difference between coating and substrate, and hence the greater the sacrificial protection offered by the coating.

Water A (2000mg/l Cl^- and 2000mg/l SO_4^{2-}) was consistently the most corrosive, with the corrosion potentials of the coating alloys generally being the lowest in this water. The high chloride addition in water B also made it more corrosive than water C which had 2000mg/l SO_4^{2-}.

No consistent changes in potential were obtained by

altering the substrate. Thus it was not possible, using potentiodynamic tests, to establish the effects of the various substrates on the performance of the coating.

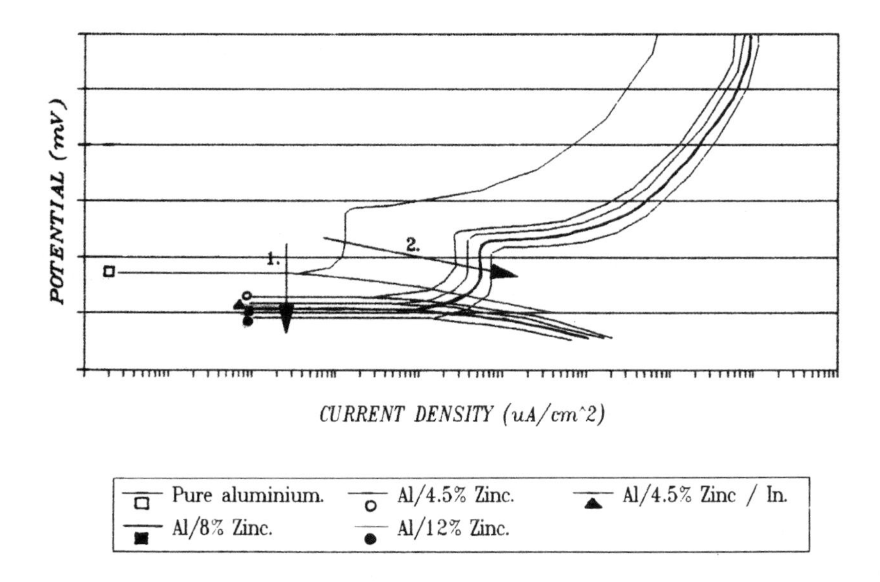

Figure 3 Trends established from the potentiodynamic tests

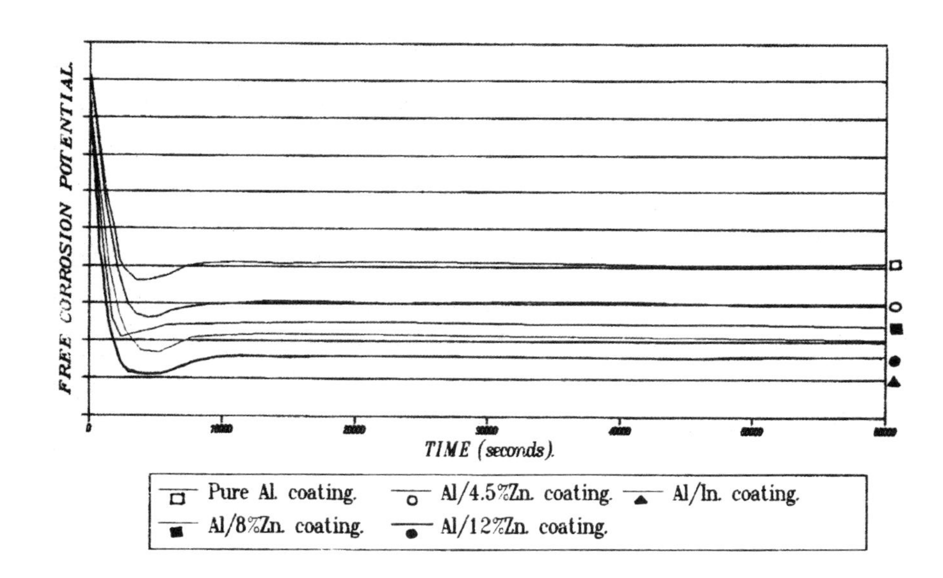

Figure 4 Trends established from the E_{corr} vs time tests

Ecorr as a Function of Time. The purpose of these tests was not merely to establish the free corrosion potential of the coating, but also to determine the time taken for the alloy to stabilize at this potential. The trends established from this work are shown in Figure 4 above.

The plots show how the coating is initially at a potential far more cathodic than its stable free corrosion potential. This may be critical if these coatings are to be used in a highly corrosive environment, where the substrate will initially corrode, sacrificially protecting the coating, which is the reverse of the desired effect.

Exposure Test Work

Salt Spray Tests. The high chloride, high sulphate water, the most corrosive of the synthetic mine waters, was selected to be used in the salt spray cabinet.

A white corrosion product started to appear on the coupon surfaces as the test continued. A close-up of this is shown in the figure below. As the photograph shows, this corrosion product is especially abundant around the defects purposely introduced in the coatings. This heavy build-up is caused by the coating sacrificially protecting the substrate, as it was designed to do. Also evident from the photograph is the fact that as the size of defect is increased, so the effectiveness of the protection decreases.

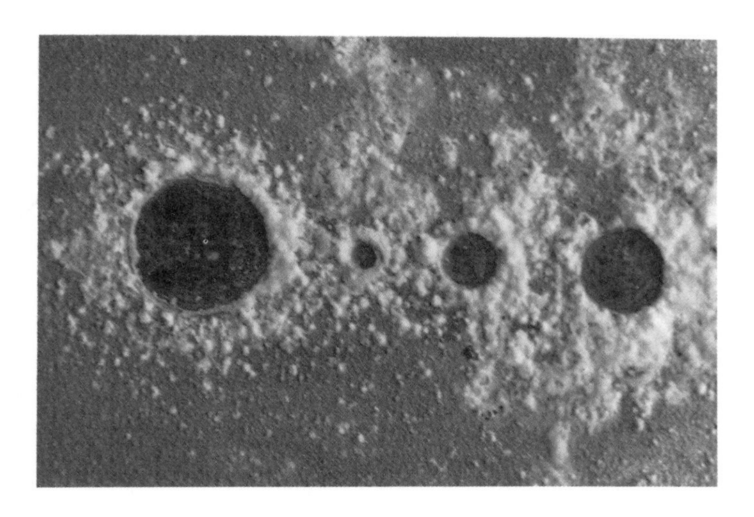

Figure 5 Close-up of a corroded salt spray coupon

After 1500 hours exposure it was apparent that the coated aluminium coupons were out-performing the coated mild steel samples. Furthermore, the best aluminium coupons were those coated with alloys having smaller zinc additions, while the largest zinc addition best protected the mild steel.

Alternate Immersion Tests. The purpose of alternate immersion was to simulate the wet/dry conditions encountered in many underground mining applications. However, exposure for periods in excess of 5000 hours still failed to corrode any of the coupons to such an extent as to allow the ranking of the various combinations. Because of this low level of corrosion a more corrosive solution, viz. 0.1N H_2SO_4, was then experimented with. After the 1000 hour sulphuric acid test, the pure aluminium coated coupons had clearly out-performed those containing zinc, with the corrosion becoming worse as the percentage zinc was increased.

Total Immersion Tests. All the total immersion coupons were suspended vertically in tanks containing the various synthetic mine waters for a period of 5000 hours. Figure 6 illustrates a typical coated mild steel coupon after this exposure period.

The regions adjacent to the holes have been corroded as shown by the black discolouration in these areas, while the rest of the coating remains completely intact. The excellent protection offered by these coatings is exemplified by the performance of the uncoated mild steel coupons, which were severely corroded after only 100 hours.

Figure 6 A coated mild steel coupon after 5000 hours exposure in a synthetic mine water

4 DISCUSSION

From the electrochemical work carried out, it is clear that as the zinc contents are increased, a decrease in potential, and hence an increase in the potential difference between coating and substrate, is produced. Furthermore the corrosion current consequently increases as the zinc content increases. These factors are of significance as they quantify the effect of zinc on the free corrosion potential and allow for the accurate measurement of the potential difference between coating and substrate. Thus, using the measured potential differences and relating them to the results obtained from the immersion trials, one could predict an optimum potential between coating and substrate. This should make it possible to formulate a coating having a specific free corrosion potential that would best protect a given substrate in a certain environment. It must be borne in mind however, that electrochemical testing alone cannot rank any of the coatings in terms of performance in any media, but can merely measure and rank the potentials at which they operate.

The salt spray test established certain trends applicable to all the mine water immersion testing. The results showed that a low percentage zinc coating is the optimum for the spraying of the aluminium alloys, including the MMC. This selection, based on the coupon performances, is justified by the fact that although the environment is highly corrosive, it is not corrosive enough to warrant any greater anodic current, due to the intrinsic corrosion resistance of the aluminium alloys. As regards the steel substrates however, a higher zinc addition, preferably in excess of 12% is recommended. The low corrosion resistance of the steel explains the accompanying high degree of sacrificial protection required.

When considering the test conducted in sulphuric acid, the reason for the superior performance of the pure aluminium coated coupons is different to that in the mine waters. The pure aluminium is far less anodic in this medium than those coatings with which zinc is alloyed. Thus, less of the coatings had corroded away during the test period. The complication with these results is that the exposed substrates on the pure aluminium coupons had completely rusted over. This suggests that even though the aluminium/zinc coatings corrode too fast, if intact, they would protect holidays in the coating far more effectively, due to their anodic nature.

From all the exposure tests conducted in the different solutions it is clear that there exists a balance between zinc additions and substrate alloy composition that will provide the optimum protection for a given substrate in a specific environment.

5 CONCLUSIONS

1. It has been confirmed that the synthetic mine waters are particularly corrosive towards mild steel, but less so towards aluminium alloys.

2. Sprayed aluminium alloy coatings on both aluminium and steel provide very good resistance to corrosion by typical mine waters, as determined by their good galvanic throwing power and their low rate of consumption.

3. It has been shown that higher-zinc coatings tend to provide the greater extent of galvanic protection required by mild steel substrates in the mine waters tested, as compared with purer aluminium coatings. Aluminium alloy substrates on the other hand, are adequately protected by pure or less zinc-alloyed coatings.

4. It is clear from the electrochemical measurements that the compositions of sprayed coatings can be tailored to achieve an optimum compromise between throwing power and coating lifetime, given the nature of the water and the type of substrate.

REFERENCES

1. V.E. Carter, 'Metallic Coatings for Corrosion Control',Newnes-Butterworth, London, 1977.
2. C. Davies and N.J. Hanford,Anti-Corrosion, July, 1970, 20.
3. B.A Shaw, A.M. Liemkuhler and P.J. Moran, ASTM STP 947,Philadelphia, 1987, 246.
4. P.O. Gartland, Mat. Performance, June, 1987, 29.
5. V.E. Carter and H.S. Campbell, J. Inst. Metals, 1960-61, 89, 472.
6. V.E. Carter and H.S.Campbell, Brit. Corr. J., 1969, 4, 15.
7. N.J. Holroyd, W. Hopples and G.M. Scamans, Int. Conf. on Fatigue, USA, 1985.
8. V. Vesely and J. Horky, Eighth Int. Thermal Spraying Conference, 1976, 430.
9. D.J. Scott, Corrosion, Dec., 1979, 111.
10. H.E. Townsend and J.C. Zoccola, Mat. Protection, 1979, 13.
11. J. Bland, American Welding Society, Florida, 1974.
12. R.A. Lyall, Eskom,s Investigations Div., Johannesburg, 1990.
13. R.M. Kain and E.A. Baker, ASTM STP 947, Philadelphia, 1987, 211.
14. A. Higginson and R.T. White, A Preliminary Survey of the Corrosivity of Water in Gold Mines, Mintek, M65, 1983.

2.2.8
Surface Prestressing by Controlled Shot Peening to Improve Resistance to Stress Corrosion Cracking

G.J. Hammersley

METAL IMPROVEMENT COMPANY INCORPORATED, DERBY, UK

1 INTRODUCTION

Stress Corrosion Cracking (SCC) is a progressive fracture mechanism in metals that is caused by the simultaneous action of a corrosive electrolyte and a sustained tensile stress.

The phenomenon was first recognised in the late 19th Century when cracks appeared, spontaneously, in the thin walled necks of brass cartridge cases used during the Indian campaigns of the British Army.

Since the problem occurred mainly in the monsoon season it became known as "Season Cracking" and was the result of SCC arising from traces of airborne ammonia, high temperatures, high humidity and a residual tensile hoop stress in the cartridge case.

Failure can be sudden and unpredictable and can occur after a few hours or many years of service. Sometimes no other form of corrosive attack is observable. Almost all alloy systems have a susceptibility to SCC, to a varying degree, under favourable conditions.

For SCC to occur a sustained tensile stress, either residual or applied, must be present.

Progressive cracking will also take place in certain circumstances under cyclic tensile loading regimes. This phenomenon is known as corrosion fatigue and although similar, is properly treated as a separate mechanism to that of SCC.

For SCC to occur four criteria must be satisfied:

a) The alloy system must be susceptible
b) The environment must be corrosive
c) Sufficient time for initiation must lapse

d) A sustained static tensile stress at the surface

An effective method of reducing or eliminating the effects of both corrosion fatigue and SCC is to induce a residual compressive stress in the surface of the material by controlled shot peening.

2 CHARACTERISTICS OF SCC

The major distinguishing characteristics of SCC may be summarised by the following:-

i) A sustained, static, tensile stress at the surface. This may be from the applied service loading or may be a residual stress due to heat treatment, welding, grinding, assembly mismatch, damage etc, but for a given combination of corrodant and alloy there is a threshold stress below which SCC will not occur.

ii) Alloys are generally more susceptible to SCC than are pure metals. In a given alloy susceptibility is highly dependent on the microstructure and therefore heat treatments or exposure to elevated temperatures in service.

iii) A given alloy will usually exhibit SCC when exposed to a small number of corrodants. It follows that chemical species which attack one alloy system may not harm others. Futhermore corrodants will only be active at certain temperatures, degrees of aeration and ionic concentration.

iv) SCC is always characterised by crack-like fissures in the surface even in ductile alloys. However in many alloys the susceptibility to SCC is greatly increased if the heat treatment has resulted in a low toughness.

v) Cracks may be inter or trans granular -or both- depending on the alloy, its microstructure, heat treatment, corrodant etc.

vi) Immunity from SCC can be conferred on susceptible alloy/corrodant combinations by inducing a residual compressive stress in the surface by controlled shot peening.

vii) Shot peening is less effective if SCC initiates at flaws or "general" corrosion pits at a depth beyond that effected by shot peening.

viii) Shot peening will delay, but not eliminate SCC in cases where, despite the induced residual stress, the applied tensile stresses reach the threshold level necessary for SCC to begin.

3 MECHANISM OF SCC

SCC has been observed in a wide variety of materials and no single explanation of the mechanism adequately covers all cases. Certainly SCC is a complex phenomenon and the basic fracture mechanisms are still a matter of some debate and conjecture, and inevitably encompass the wider question relating to the "borderline" between crack initiation and propagation.

It is not the purpose of this paper to enter into this debate, but to take a more pragmatic approach in analysing the mechanisms which certainly do occur and more to the point how these are affected by controlled shot-peening.

All SCC mechanisms can be broadly categorised into two, viz., Anodic and Cathodic Mechanisms.

Anodic Mechanism

This simply involves dissolution and removal of material from the crack tip.

Most metals have the ability to form an oxide layer at the surface which, though very thin, offers a protective barrier against corrosion.

If this layer becomes ruptured there is an electrochemical potential difference between the newly exposed "active" metal surface (the Anode) and the passive film (the cathode). The relatively small area of the anode (the crack tip) compared with that of the cathode (the rest of the surface) results in an adverse anode to cathode ratio and high material dissolution rates will occur at the anode (Anodic Dissolution).

If the freshly exposed metal surface can form a new passive layer faster than a new surface is created by rupture then the mechanism will stop.

However, chemical attack - such as by chloride ions - or surface rupture caused by tensile stress will continue to drive the mechanism.

Anodic dissolution is also the result of slight differences in material composition - since no material is completely homogeneous - or variations in internal strain. These also result in electrochemical potential differences.

The Anodic Mechanism appears to be the dominant driver of SCC in copper alloys and monel and is responsible for intergranular SCC of some austenitic stainless steels.

Cathodic Mechanism

This often begins with the Anodic Mechanism, since as a natural consequence of the electrolytic anodic dissolution there is an increase in the concentration of hydrogen ions, and hence an increase in the pH of the solution within the crevice. The hydrogen ions pick up electrons from the environment to produce hydrogen gas which, in turn, is diffused into the metal and results in hydrogen embrittlement.

The Cathodic Mechanism - or hydrogen induced crack growth - is the phenomenon principally responsible for SCC in ferritic steels and also in nickel base alloys and the alloys of Aluminium and Titanium.

Regardless of the dominant SCC mechanism, the material of construction or the corrosive environment, the inducement of a residual compressive stress, in the surface of the metal, by shot peening can be an effective measure against SCC.

4 SUSCEPTIBLE ALLOY SYSTEM/CORRODANT COMBINATIONS

Many lists have been drawn up to relate corrodants and alloy systems which are known to be susceptible combinations to SCC. New combinations are being added all the time and there appears to be no limit to the possibilities. A good guide is shown in Table 1. [1]

5 THE SHOT PEENING PROCESS

Shot peening is a cold working process which involves the bombardment of metal components with spherical media - or shot. Each impact site is evidenced by an indentation or dimple at the surface. For this to occur it follows that the surface has yielded whilst the material in the core of the component exerts an opposite reaction thus inducing a hemispherical field of compressive residual stress in the surface below the dimple. (Figure 1)

As the complete surface becomes covered with the small impact craters then the dimples overlap and a compressive layer of even depth is generated over the surface of the whole component.

Table 1 **Some environment-alloy combinations known to result in SCC**

Environment	Alloy system								
	Aluminum alloys	Carbon steels	Copper alloys	Nickel alloys	Stainless steels austenitic	Stainless steels duplex	Stainless steels martensitic	Titanium alloys	Zirconium alloys
Amines, aqueous	...	●	●	...	...	...	...	...	...
Ammonia, anhydrous	...	●	...	...	...	...	...	...	...
Ammonia, aqueous	...	...	●	...	...	...	...	...	...
Bromine	...	...	...	...	...	...	...	...	●
Carbonates, aqueous	...	●	...	...	...	...	...	...	...
Carbon monoxide, carbon dioxide, water mixture	...	●	...	...	...	...	...	...	...
Chlorides, aqueous	●	...	...	●	●	●	...	...	●
Chlorides, concentrated, boiling	...	...	...	●	●	●	...	...	...
Chlorides, dry, hot	...	...	...	●	...	...	...	●	...
Chlorinated solvents	...	...	...	...	...	...	...	●	●
Cyanides, aqueous, acidified	...	●	...	...	...	...	...	...	...
Fluorides, aqueous	...	...	...	●	...	...	...	...	...
Hydrochloric acid	...	...	...	...	...	...	...	●	...
Hydrofluoric acid	...	...	...	●	...	...	...	...	...
Hydroxides, aqueous	...	●	...	...	●	●	●	...	...
Hydroxides, concentrated, hot	...	...	...	●	●	●	●	...	...
Methanol plus halides	...	...	...	...	...	...	...	●	●
Nitrates, aqueous	...	●	●	...	...	...	●	...	...
Nitric acid, concentrated	...	...	...	...	...	...	...	...	●
Nitric acid, fuming	...	...	...	...	...	...	...	●	...
Nitrites, aqueous	...	...	●	...	...	...	...	...	...
Nitrogen tetroxide	...	...	...	...	...	...	...	●	...
Polythionic acids	...	...	...	●	●	...	...	...	...
Steam	...	...	●	...	...	...	...	...	...
Sulfides plus chlorides, aqueous	...	...	...	...	●	●	●	...	...
Sulfurous acid	...	...	...	...	●	...	...	...	...
Water, high-purity, hot	●	...	...	●	...	...	...	...	...

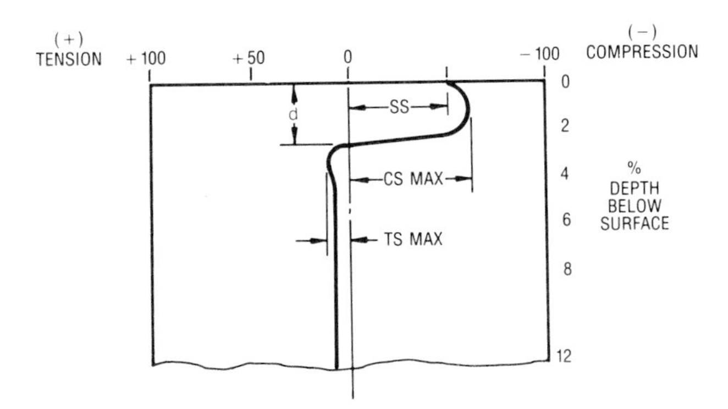

Figure 1 Example of residual stress profile created by shot peening

Since cracks cannot propagate through a zone of compressive stress the process is very valuable in eliminating a wide range of fatigue and stress related failure mechanisms. Most commonly the process is used to enhance the life of components which are subjected to bending, and it follows that the greatest benefit is bestowed on those parts that suffer the greatest surface strains, such as springs, gear teeth, reciprocating parts in internal combustion engines and most components in gas turbine engines.

In addition to inducing residual compressive stress benefits may also flow from surface hardening (work hardening effect), refinement of the grain structure and metallurgical transformation such as the conversion of retained austenite to martensite in case hardened steels.[2]

Shot peening parameters and controls

The most important factor in a shot peening "design" is to achieve the optimum residual stress profile. This will vary with the type of alloy, dimensions and service conditions.

The depth of induced compressive stress is the same dimension as the diameter measured across the dimple on the surface, and the peak magnitude is about 60% of the Ultimate Tensile Stress (UTS) of the material. Since the shot peened component is in equilibrium it follows that a balancing core tensile stress must be present.

Since the depth of induced residual compressive stress is the same as the diameter across an impact dimple, the optimum stress profile for a given component can be determined provided that its surface hardness is known. see Figures 2 & 3

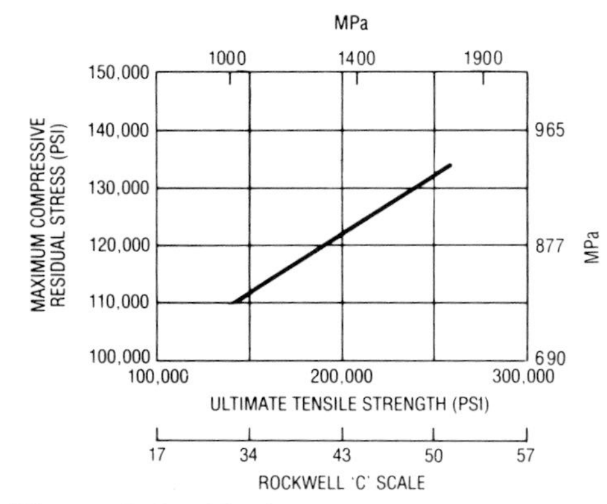

Figure 2 Residual stress produced shot-peening vs. tensile strength of steel

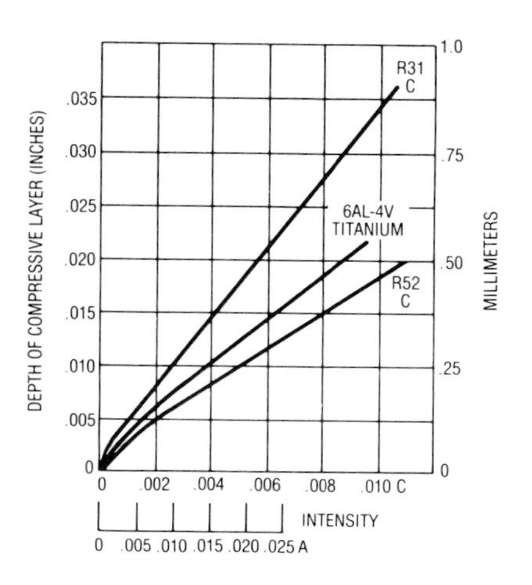

Figure 3 Depth of compression vs. Almen arc height

The magnitude and depth of the residual compressive stress are affected by the following properties of the shot peening media:-

Material
Hardness
Density
Velocity
Direction (angle of impingement)
Size
Shape

All of these properties are inter-related and can be measured very accurately as a single parameter; viz., Intensity.

Using the property of stretching by cold work, thin test strips of spring steel, of closely controlled composition, treatment and dimensions are peened on one side only - the other being held against a rigid steel block. On removal from the block, since only one side has been stretched the test strip exhibits a curvature - convex in the direction from which the shot was fired. The degree of curvature is a direct measure of intensity and can be related to the desired stress profile. The test strips are known as Almen strips after the inventor of this system for the determination of shot peening intensity. (Figure 4)

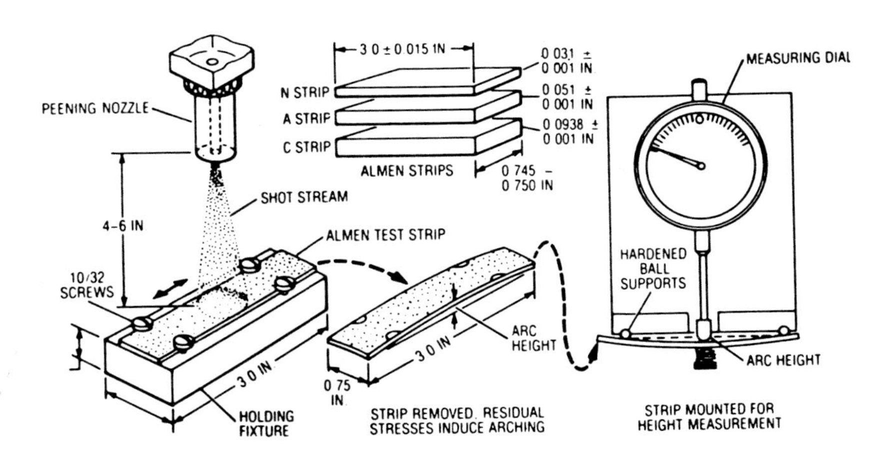

Figure 4 The Almen strip system

Since the induced compressive stress must be of an even depth and magnitude over the entire surface of the component, the degree of "coverage" of the surface is also an important parameter. 100% coverage is defined as "complete obliteration of the original surface".

Determination of coverage is a little more subjective a measurement and requires some skill and experience on the part of the observer. The use of fluorescent dyes has greatly improved the determination of coverage and is a method which is indispensible on components which require peening over large areas.

Since, in addition to inducing a residual compressive stress, shot peening can, in some materials, increase the surface hardness, refine the grain structure or change the metallurgy, then extended coverage can be beneficial, and coverage rates of 200% and 400% are often specified.

6 EVALUATION OF SHOT PEENING EFFECTS ON SCC

The challenge to the SCC investigator is that while it is relatively easy to determine whether an alloy is susceptible to SCC , it is far more difficult to determine if it possesses a degree of susceptibility which will restrict its general usefulness.

A wide variety of laboratory tests has been developed to determine susceptibility to SCC and therefore to evaluate the effectiveness of treatments for it. (Figure 5)

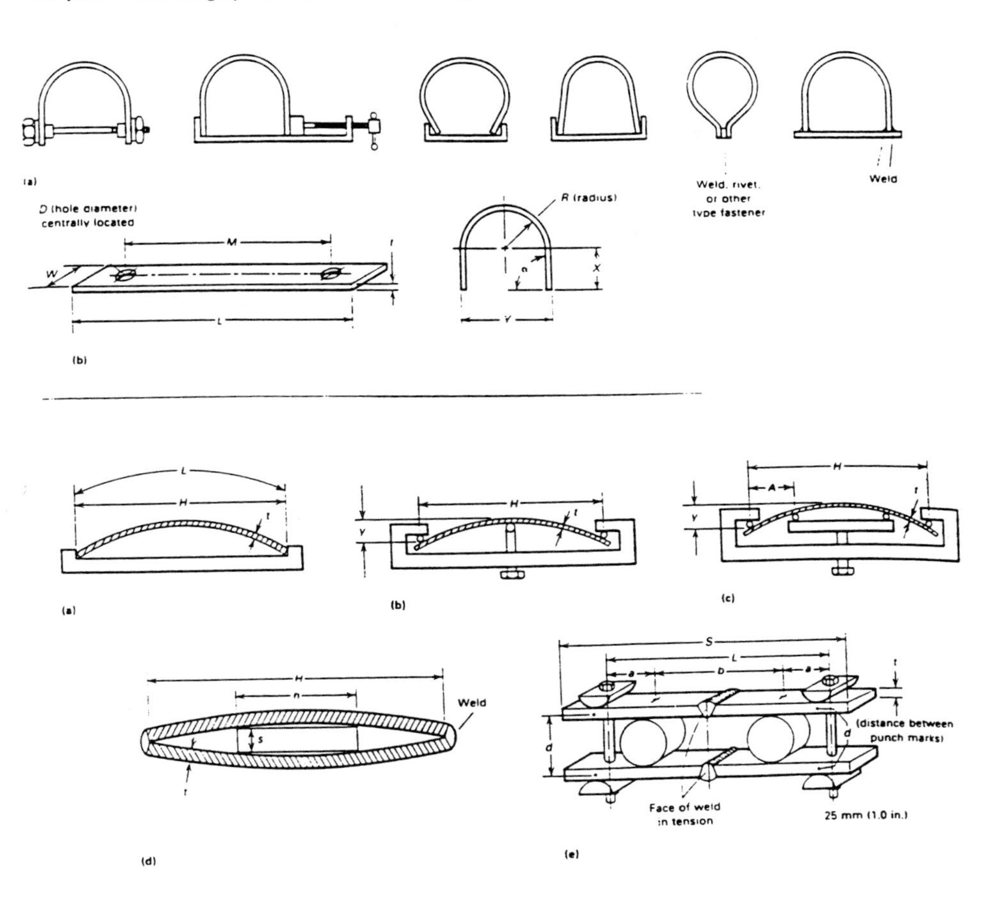

Figure 5 Examples of "U-bend" and "bent beam" SCC test specimens

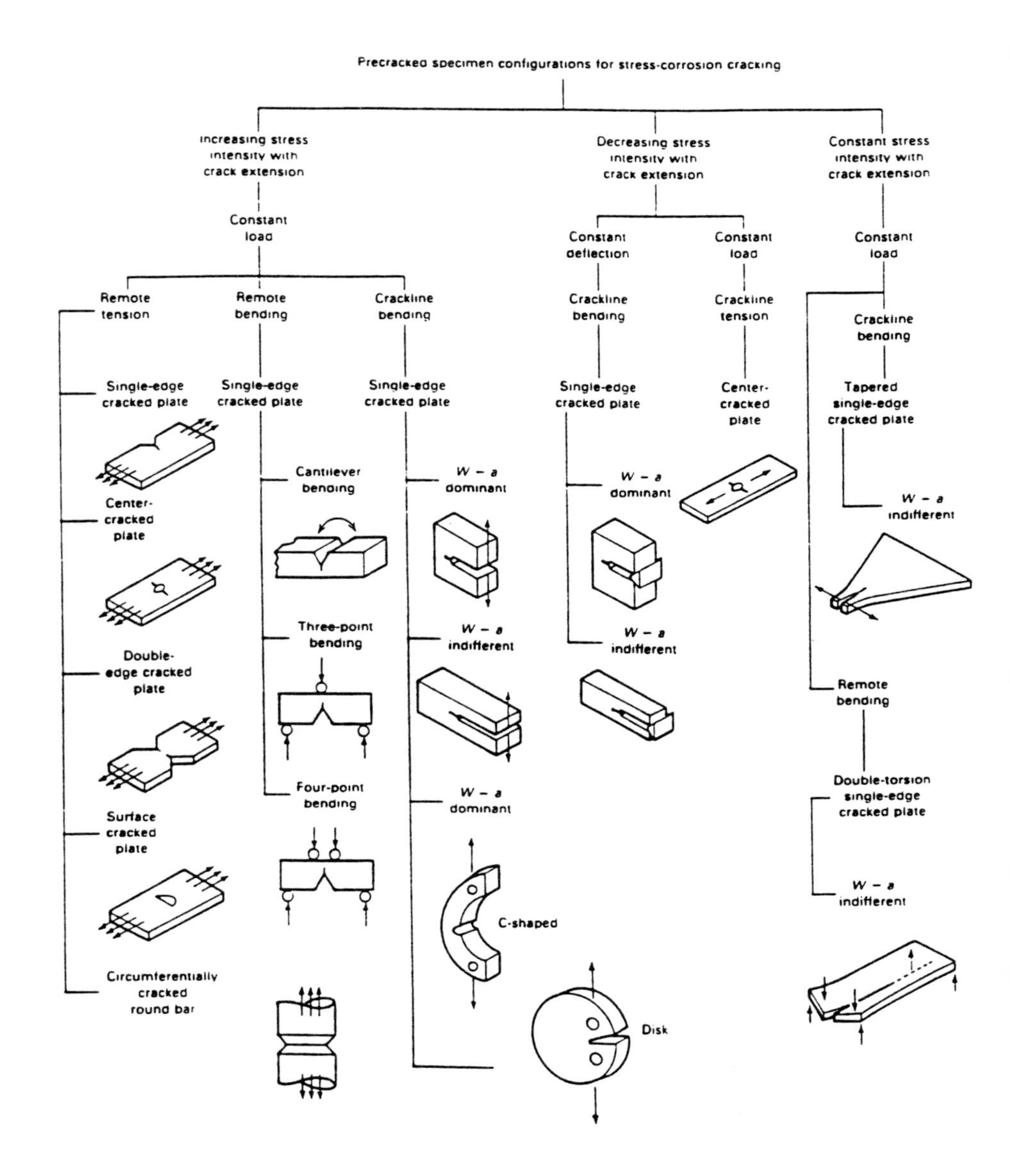

Figure 6 Precracked specimen configurations for stress corrosion cracking

The most widely used, and easiest to perform, category of tests combine crack initiation and propagation. (Figure 6)

In these, specimens are strained and exposed to a corrosive environment. A wide variety of samples, methods of applying the static stress, and environmental exposure regimes exist and range from severe conditions to determine quickly whether a material is susceptible to SCC or measure the relative effect of preventative means - such as coatings or controlled shot peening.

The effects of shot peening are most graphically illustrated in figure 7 which shows cracking on an untreated sample after 2 hours and no cracking on a shot peened sample of after 1000 hours. The tests involved immersion of "U bend" samples of 304 stainless steel in boiling 42% Magnesium Chloride.

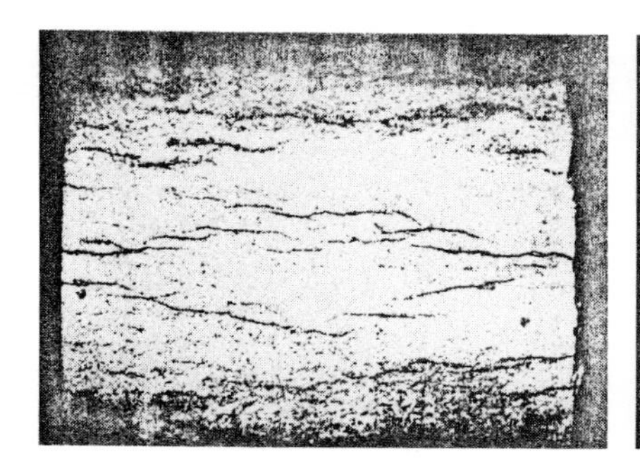

8X

a. Unpeened

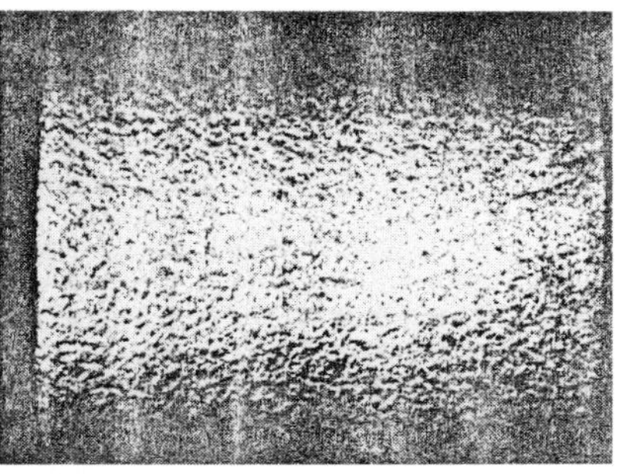

8X

b. Shot Peened

Figure 7 Type 304 stainless steel U-bend specimens after stress corrosion tests in boiling 42% magnesium chloride

Although effective against SCC in a wide range of alloy/corrodant systems, shot peening is most commonly used to combat the phenomenon in austenitic stainless steels, for the following reasons.

a) Austenitic stainless steels can be very susceptible to SCC.
b) The materials are in wide use in corrosive environments - particularly as they are frequently specified in the belief that they are indeed "stainless" or incapable of corrosion.
c) Many of the components are vessels of welded structure which have very high residual stresses in the heat affected zones (HAZs).
d) Austenitic Stainless Steels respond very well to the shot peening process.

e) Intergranular corrosion can be prevented by shot peening prior to exposure to sensitizing temperature [4, 5]. see Figure 8.

A comparison of the effectiveness of treatments against SCC in steels [3] is given in Table 2.

Table 2

Effectiveness of SCC Prevention Methods

Prevention Method	Ferritic	Austenitic	Martensitic	Duplex
Materials Selection	1,2,3	1,2,3	1,2,3	1,2,3
Barrier Coatings	3,	3	4	3
Eliminating Corrodant	1,2	1,2	1,2	1,2
Adding Inhibitors	N/A	2	N/A	N/A
Thermal stress relief	4	4	X	4
Heat Treatment	*	*	1,3	X
Shot Peening	1,2,3	1,2,3	4	1,2,3
Cathodic Protection	XX	1	XX	1
Lowering the temp	2,3	2,3	X	2,3
Design Techniques	2	2	2	2

LEGEND

1. Eliminates SCC in processes having high corrodant concentration.

2. Eliminates SCC in processes having concentration effects (cyclic drying & wetting).
3. Eliminates external cracking from atmospheric SCC corrodants.

4. Delays onset of SCC.

N/A No industrial applications known.

X Not effective.
XX Not effective - may accelerate cracking.
* Effective only against intergranular SCC of sensitized material.

7 EXAMPLES OF SCC IN SPECIFIC INDUSTRIES & ENVIRONMENTS

SCC is not by any means confined to chemical process industries. Indeed hardly any alloy system in use in any industry is immune to the phenomenon. The

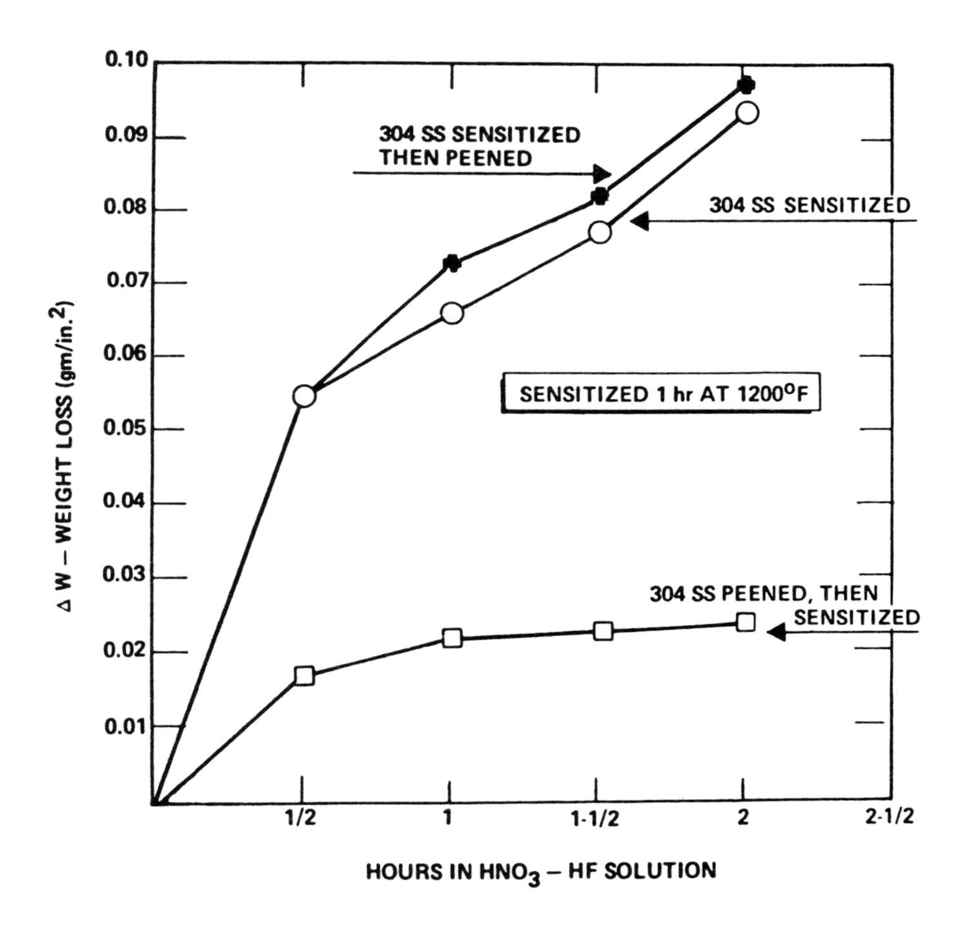

Figure 8 Intergranular corrosion of peened and unpeened Type 304 stainless steel

following is a listing by industry of some of the occurrences of SCC which have been successfully controlled by shot peening.

Aircraft

1. Airframe structural members manufactured from the "2 and 7 - thousand" series of high strength aluminium alloys.

2. Undercarriage parts and fasteners in ultra - high strength steels such as 300m and H-11.

3. Tubing & tube fittings in precipitation hardened stainless steels such as AM350 and 17-4PH.

Chemical

1. Tanks, vessels and piping in austenitic stainless steels such as types 304, 316, 321 & 347.

2. Storage vessels of carbon and low alloy steels for anhydrous ammonia.

3. Zirconium lined pressure vessels.

4. Copper-silicon alloy tanks for sulphuric and storage.

Petroleum and Petrochemical

1. Carbon steel drill pipes and casings (sulphide SCC attack).

2. Heat exchangers and vessels (Caustic SCC attack).

3. Carbon steel flash vessels and associated pipework exposed to aqueous amine SCC attack.

4. Refining units in type 300 series stainless steels (chloride SCC).

5. Heat exchanger tubes of copper-zinc alloys (aqueous ammonia SCC).

6. Hydro-treating units in unstabilized (304) austenitic stainless steel (Polythionic acid SCC).

7. High strength steel bolting and compressor rotors (Hydrogen sulphide SCC).

Nuclear Power Plants

1. Intergranular SCC in sensitised type 304 stainless steel piping in boiling water reactors and SCC in nickel based steam generator tubing.

2. Intergranular SCC in nickel base alloy jet pump beams.

3. SCC in LP steam turbine rotors in chrominium/molybdenum/vanadium steels.

Fossil Fuel Powerplant

1. SCC in brass condenser units.

2. Chloride SCC in type 304 stainless steel feedwater heaters.

3. Carbon steel de-aerators.

4. Austenitic stainless superheater tubing.

5. LP steam and gas turbine blades and discs in martensitic stainless steels.

8 CONCLUSIONS

Stress Corrosion Cracking is a much more common phenomenon than is widely appreciated.

The induction of a residual compressive stress by controlled shot peening can be an effective remedy - or delay the onset of SCC [6,7,8].

Careful selection and control of the parameters of shot peening is necessary to ensure a consistent degree of induced surface pre-stress and hence effective, all-over, treatment against SCC.

REFERENCES

1. Metals Handbook, 9th edition, volume 12, "Corrosion", ASM International.

2. Yashio Okada et al., "Development of High Strength Transmission Gears", Society of Automotive Engineers Technical Paper, 1992.

3. Dale R. McIntyre, "How to prevent Stress Corrosion Cracking in Stainless Steels I & II", Battelle Memorial Institute, April 7th 1980.

4. US Patent No. 3,844,846.

5. W. H. Friske, "Shot Peening to Prevent the Corrosion of Austenitic Stainless Steels", Rockwell International Corporation, 1975.

6. J. H. Milo, "Prevention of Stress Corrosion Cracking by Shot Peening", National Association of Corrosion Engineers, 1968.

7. "Minimising Stress Corrosion Cracking in Heat Treatable Wrought Low Alloy and Martensitic Corrosion Resistant Steels", Society of Automotive Engineers, Aerospace Recommended Practice No. 1110, 1969.

8. Bob Gillespie, "Controlled Shot-Peening can help prevent Stress Corrosion Cracking", Proceedings:- Deutsche Gesellschaft Fur Metallkunde.

Section 2.3 Heat Engines

2.3.1
High Temperature Diffusion Barrier Coatings

O. Knotek, F. Löffler, and W. Beele

AACHEN UNIVERSITY OF TECHNOLOGY, INSTITUT FUER WERKSTOFFKUNDE B, TEMPLERGRABEN 55, W-5100 AACHEN, GERMANY

1 INTRODUCTION

A main aim in today's development of aero gas turbines is higher efficiency combustion. One possibility to optimize the combustion is an increase of the temperature level in the whole turbine system. This will be limited by material interactions appearing at higher temperatures. Actual Ni-base turbine blades are corrosion protected by a sprayed MCrAlY overlay and a ceramic thermal barrier top layer. This system offers a long lifetime when it is used at temperatures below 1000^{0}C in the substrate interface zone.

For the next turbine generation, interface temperatures above 1100^{0}C between the MCrAlY layer and the Ni-base turbine blade are demanded. At this temperature level, interdiffusion between these materials takes place and rapidly destroys the interface zone by changing structures and growing Kirkendahl voids[1,2]. Therefore a diffusion barrier coating, deposited by the PVD sputter process, is in development.

2 THE Ti-Al-O-N COATING SYSTEM

The Ti-Al-O-N system offers many different layer structures. Depending on the elemental and content three important types of structure are achievable:

1. Columnar grown TiN-cells with a substitution of Ti by Al plus Ti-oxides.
2. A two phase coating with the fcc-TiN structure and alumina structures.
3. Amorphous alumina with thermal stabilization by the addition of interposed nitrogen[3].

The amorphous type (No.3) has been suggested as probably the best structure design for diffusion barrier properties because it is well know that diffusion of elements appears along "hot paths" like structure faults and grain boundaries[4]. On the other hand it was reported in the literature that the mechanical properties of pure Al-O-N coatings are in comparison to alumina lower[3], so it was decided to investigate a coating with an added Ti-content for improved mechanical behaviour. These decisions were taken from the view point that the cyclic stress peaks appearing on aero foils (thermal stresses and creep stresses) are very high. A wide range of variation

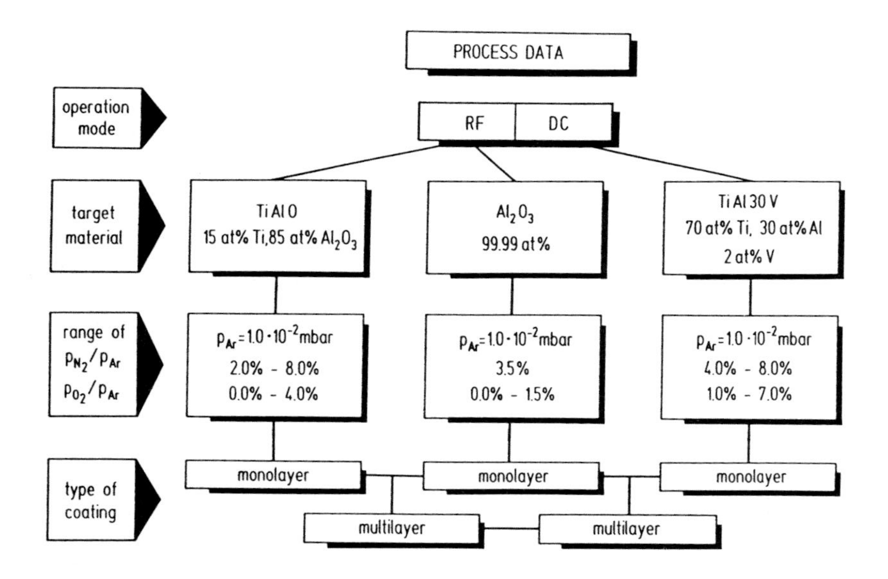

Figure 1 Process data for the deposited Ti-Al-O-N coatings

was investigated using three different sputter targets with several gas compositions in the vacuum chamber during deposition. Figure 1 gives an overview of the field of deposited coatings in the Ti-Al-O-N system.

X-ray Structure Examination

The coating structure was examined by an X-ray diffractometer. The change in structure was examined by comparing the structure directly after the deposition and after thermal treatments in a vacuum furnace. While all pure Al-O-N coatings were in all steps of investigation X-ray amorphous, most of the Ti-containing coatings yielded crystalline structures (see Figure 2).

X - ray diffractometry												
target: TiAlO (15at% Ti, 85at% Al_2O_3)							target: TiAl30V (≈69at% Ti, ≈29at% Al, <2at% V)					
substrate: CMSX-6							substrate: CMSX-6					
unannealed	Ti_2N δ-Al_2O_3	X-ray amorphous	X-ray amorphous	X-ray amorphous	TiN Ti_2N	TiN	X-ray amorphous	X-ray amorphous	Ti_2O β-Al_2O_3 δ-Al_2O_3	Ti_2N α-Al_2O_3 δ-Al_2O_3	Ti_2O	α-Al_2O_3 δ-Al_2O_3
900°/3h					TiN Ti_2N	Ti_2O	X-ray amorphous	Ti_2O δ-Al_2O_3	Ti_2O δ-Al_2O_3		β-Al_2O_3 δ-Al_2O_3 η-Al_2O_3 κ-Al_2O_3 Ti_2O	α-Al_2O_3 δ-Al_2O_3
1100°C/3h	δ-Al_2O_3	X-ray amorphous	X-ray amorphous	α-Al_2O_3	TiN Ti_2N	Ti_2O	X-ray amorphous	Ti_2O β-Al_2O_3 δ-Al_2O_3	Ti_2O δ-Al_2O_3	α-Al_2O_3	δ-Al_2O_3 Ti_2O	α-Al_2O_3 δ-Al_2O_3
partial pressure ratio p_{N_2}/p_{Ar}	2	4	6	4	8	6	3	5	5	6	6 %	8
partial pressure ratio p_{O_2}/p_{Ar}	0	0	0	1.5	1.5	4	7	5	7	1.5	2.7 %	1.5
reactive gas pressure ratio p_{reac}/p_{Ar}	2	4	6	6.5	9.5	10	10	10	12	7.5	8.7 %	9.5

Figure 2 Coating structure of various Ti-Al-O-N compositions before and after thermal treatment

The Ti-rich coatings (target with a Ti-content of 70at.%) were X-ray amorphous or quasi-amorphous only under conditions of low P_{N2} and high P_{02}. With higher nitrogen content and $P_{N2}/P_{02} > 2$, TiN or alumina phases always appeared. After annealing, the intensity of the peaks very often increased which indicated that the columnar structures grew at high temperature.

With low Ti-contents (target with 15 at.% Ti, 85 at.% Al_2O_3) it was possible to deposit X-ray amorphous films if P_{N2} was in the correct range. If the layer was amorphously deposited it remained amorphous after annealing at 1100^0C for 3 hours. The range was also dependent on the nitrogen and oxygen pressures. The nitrogen had to be between 2 and 6% P_{N2}/P_{Ar} while no oxygen was added. With oxygen it was impossible to make amorphous coatings.

In summary, all Ti-Al-O-N coatings were not as good as the stabilized coatings which was indicated by the very small ranges in which it was possible to deposit amorphous coatings and by the recrystallisation processes of most of the compositions.

3 INTERDIFFUSION BETWEEN MCrAlY AND CMSX 6

High performance superalloys for aero foils are cast as single crystal material. The CMSX 6 (Cannon-Muskegon Single Crystal No. 6) is one of these low density superalloys. A conventional MCrAlY overlay (MTS 1262B) was sprayed on CMSX 6 specimens. Table 1 gives the nominal compositions of both materials.

Figure 3 shows how the interface structure is changed by interdiffusion. After 1100^0C/100h in a vacuum furnace, the interface is perforated by Kirkendahl voids. Below the void zone, γ'-phases are formed. These brittle zones are ideal for crack initiation during the mechanical stress cycling in a turbine.

Table 1 Chemical composition of CMSX 6 and MTS 1262B (origin: MTU; H.C. Starck)

	CMSX 6	MTS1262B
Ni [wt%]	balance	32.5
Co [wt%]	5.0	balance
Cr [wt%]	9.8	20.9
Mo [wt%]	3.0	-
Ta [wt%]	2.0	-
Al [wt%]	4.8	7.5
Ti [wt%]	4.7	-
Hf [wt%]	0.1	-
Y [wt%]	-	0.4
C [wt%]	-0.006	0.02

Void formation was clearly detected after static thermal treatment. The micrographs shown in Figures 3 (c) and (d) demonstrate that the interface zone is changed uniformly. The darker grey coloured γ'-phases are easily detectable.

Figure 4 shows structure etched cross-sections where the size of the diffusion zone after the 100h treatment is illustrated. The diffusion directions of the responsible elements were examined by an analysis of the chemical compositions before and after the thermal treatment.

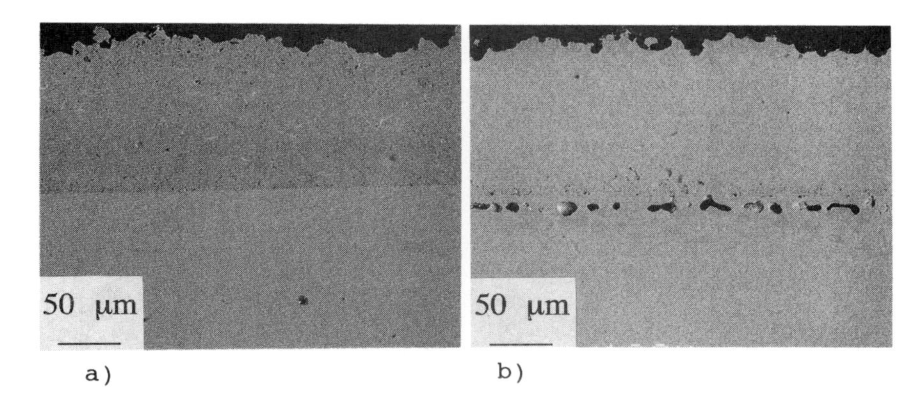

Figure 3 SEM micrographs of polished cross-sections of MCrAlY on CMSX 6: (a) as-sprayed, (b) after 1100°C/100h

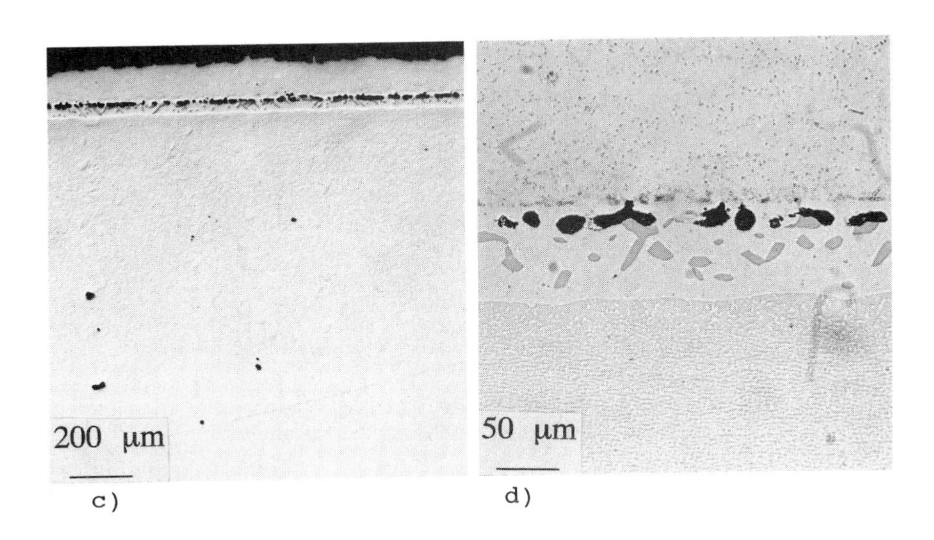

Figure 3 Micrographs of polished cross-sections of MCrAlY on CMSX 6 afer 1100°C/100h

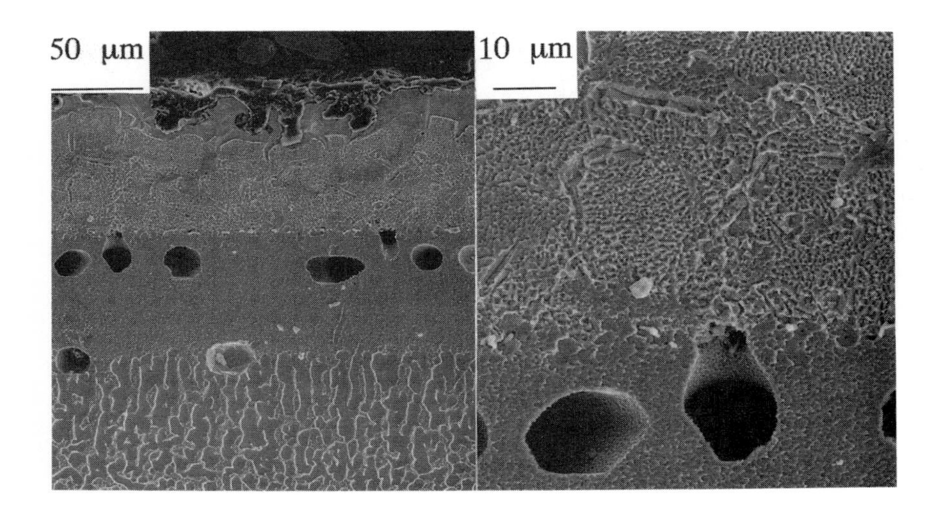

Figure 4 SEM micrographs of a structure etched MCrAlY-CMSX 6 cross-section after $1100^0C/100h$; 8 s etched with 150 ml HCl, 50 ml H_2O, 25 g CrO_3

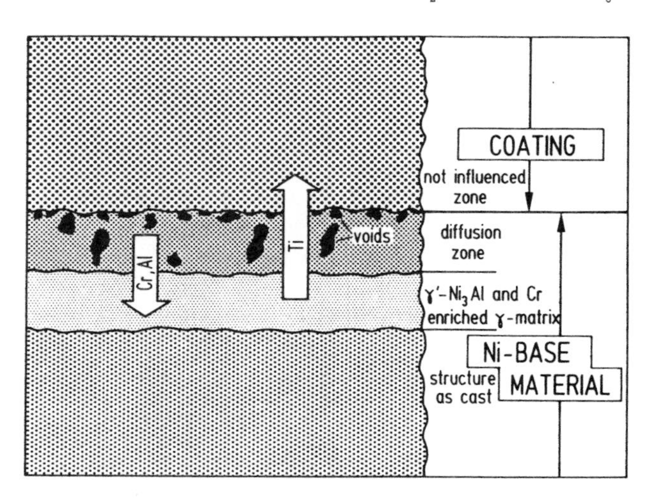

Figure 5 Diffusion directions in the system MCrAlY-CMSX 6 at 1100^0C

Figure 5 gives a schematic of the general diffusion directions. Cr, Al, and Co are diffusing into the Ni-based substrate while Ti is diffusing into and through the MCrAlY.

4 INTERDIFFUSION WITH A BARRIER COATING

Besides the reduction of diffusion during the whole lifetime, the diffusion barrier had to be designed in a way that a starting diffusion should be available for a good adhesive bond. On the other hand the coating has to be defect-free after typical stress cycling and the properties of the whole system (hot gas corrosion and temperature gradient through the surface) should not be influenced. Therefore one restriction to the barrier design was that the barrier has to be as thin as possible (see Figure 6).

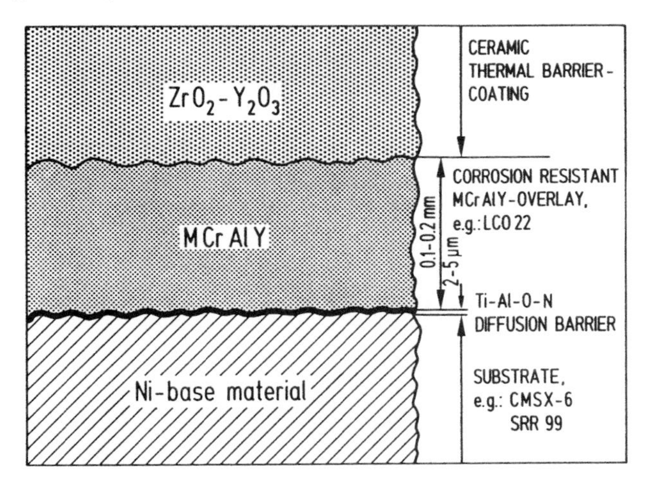

Figure 6 Schematic of the thickness profiles of the coating system including the diffusion barrier

As indicated by the X-ray diffractometry, the Ti-containing diffusion barriers showed no worthwhile improvement. On one hand this was caused by the unstable crystal structure, on the other hand it was determined that especially the Ti from the barrier reacted with the MCrAlY in such a way that the barrier layer became porous. This effect is shown in Figure 7.

The Al-O-N barrier coatings showed promising behaviour. With their amorphous structure, stabilized up to 1100^{0}C, they are able to act as a passive working barrier as illustrated in Figure 8. In the polished condition no diffusion zone was detected. Only a small zone above the coating showed any bond reaction. The structure etching of the cross-section showed that a

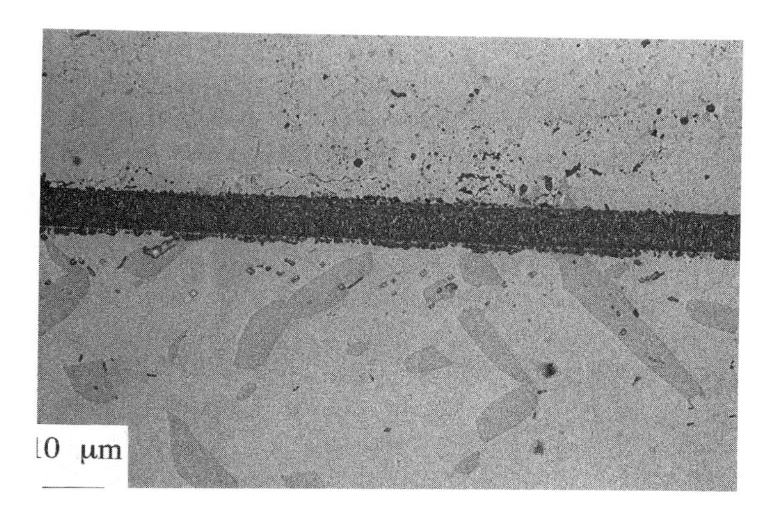

Figure 7 Cross-section of a Ti-Al-O-N coating after $1100^0C/100h$

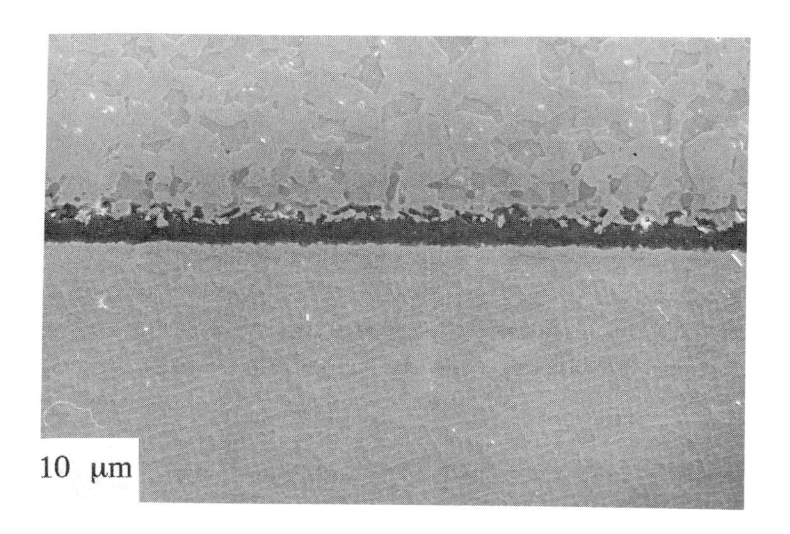

Figure 8 Cross-section with an Al-O-N barrier after $1100^0C/100h$

diffusion reaction had taken place (see Figure 9). This starting diffusion gave reasonable adhesion as determined by thermal cycling between 200^0C and 1100^0C.

The next step in the examination was a treatment up to $1100^0C/400h$ including thermal cycling as mentioned above. The Al-O-N barrier also showed excellent

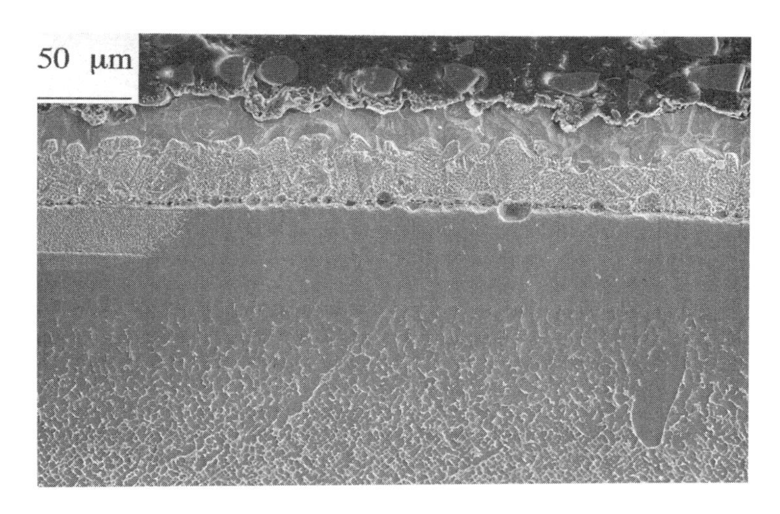

Figure 9 Etched cross-section of an Al-O-N barrier after 1100°C/100h; 8 s etched with 150 ml HCl, 50 ml H_2O, 25 g CrO_3

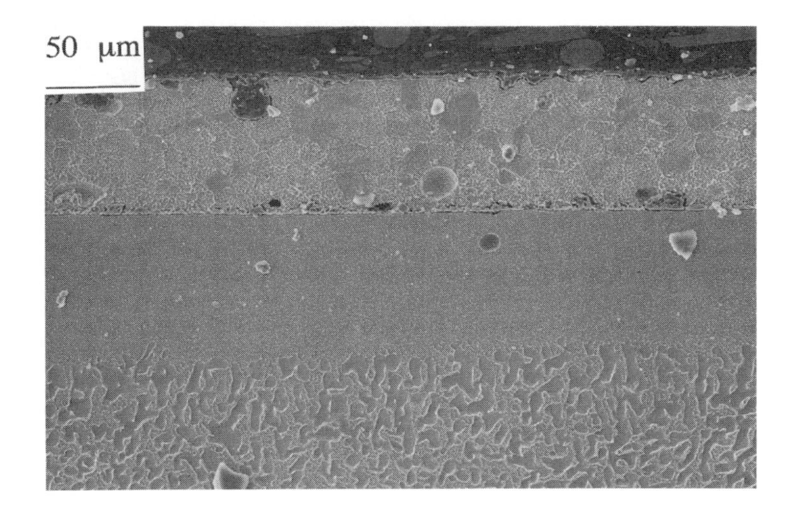

Figure 10 Etched cross-section of an Al-O-N barrier after 1100°C/400h plus 20 thermal cycles; 8 s etched with 150 ml HCl, 50 ml H_2O, 25 g CrO_3

performance after this test. As shown in Figure 10, the size of the diffusion zone was still comparable to the state after 100h. Kirkendahl voids are still not detectable.

Only some local defects were detected. These defect zones showed the typical diffusion behaviour of the system without a barrier coating. The example given in Figure 11 shows such a defect area on the right half after a 100h treatment. It is still not clear if this defect type is created during the thermal treatment or if it is caused by the spraying process on the PVD coating.

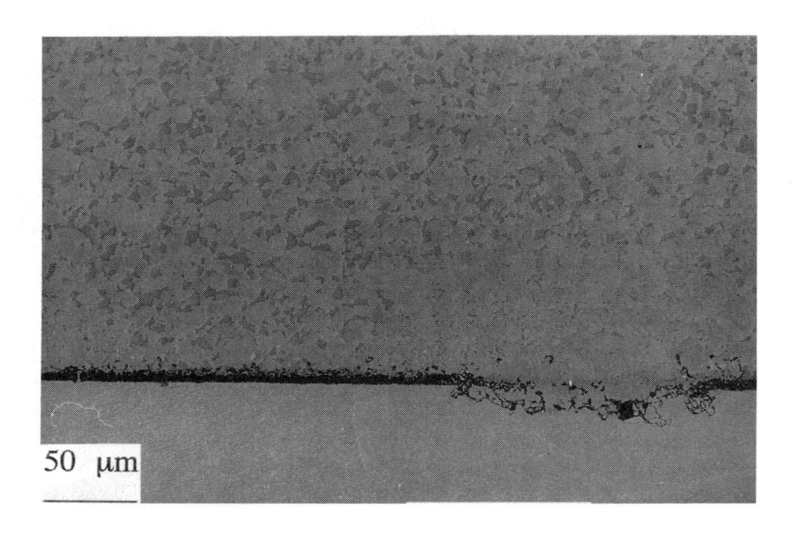

Figure 11 Cross-section of an Al-O-N barrier after 1100^{0}C/100h; left half still intact, right half with diffusion

5 CONCLUSIONS

By investigating several coating compositions in the material system Ti-Al-O-N it has been shown that a stabilized Al-O-N coating offers very good barrier properties. Annealing treatments up to 400h at 1100^{0}C showed that good protection against interdiffusion is available with only 1 to 1.5μm thick barrier coatings.

The success of this kind of barrier depends on the quantity and distribution of defects through the coating. Additional tests have to determine how these defects can be minimized. The fracture toughness of the whole system has to be assessed by creep tests and thermal fatigue tests on real turbine blades. It has to be proved how far the whole deposition process can be optimized, e.g. by deposition of the barrier and the MCrAlY "in one go".

ACKNOWLEDGEMENTS

The authors would like to thank Dr. Peichl, MTU München, for the helpful discussion and Mr. Platz, MTU München, for carrying out the MCrAlY spraying.

REFERENCES

1. J.R. Rairden, Thin Solid Films, 1978, 53, 251.
2. R. Pichoir and J.E. Restall, 'Protective Coatings for Superalloys in Gas Turbines', ONERA, 1984, 12.
3. T. Leyendecker, 'Über neuartige Schneidwerkzeug-beschichtungen auf Titan- und Aluminiumbasis', Doctoral Thesis, Aachen Univ. of Technology, 1985.
4. H. Kaesche, 'Die Korrosion der Metalle', Springer Verlag, 1990.

2.3.2
The Degradation of ZrO_2–Y_2O_3 Thermal Barrier Coatings by Molten/Semi-molten Middle East Sand

D.J. de Wet, F.H. Stott,[1] and R. Taylor[2]

DIVISION OF MATERIALS SCIENCE AND TECHNOLOGY, CSIR, PO BOX 395, PRETORIA, 0001, SOUTH AFRICA

[1] CORROSION AND PROTECTION CENTRE, UMIST, PO BOX 88, MANCHESTER M60 1QD, UK

[2] MATERIALS SCIENCE CENTRE, UMIST, GROSVENOR STREET, MANCHESTER M1 7HS, UK

1 INTRODUCTION

The continuous demand for increased performance and fuel efficiency of turbine engines, has resulted in the continual increase in turbine operating temperatures. This has led to the development of single crystal blade technology, advanced MCrAlY protective coatings, and sophisticated cooling techniques. The introduction of ceramic thermal barrier coatings, which exhibit a low thermal conductivity, offers considerable potential for significantly improving gas turbine performance and durability.[1]

During turbine operation on Middle-East routes, the high pressure turbine blades of several civil engines exhibited deposition of abnormal contaminants on the aerofoil pressure surfaces (Figure 1).[2,3] The deposits consisted of two apparent types. The first was a complex calcium-magnesium silicate, with a structure relating closely to the mineral diopside ($CaMg(SiO_3)_2$), while the second was predominantly anhydride calcium sulphate ($CaSO_4$), originating from the injection of gypsum ($CaSO_4.2H_2O$) as mineral dust.[2]

Both the detected phases resulted from ingestion of mineral debris unique to the Middle-East area during take-off and landing. Fusion and subsequent chemical reaction occurred within the combustion chamber to produce the observed complex silicate. Following the formation of the silicate in the combustion chamber, the compound initially existed as a liquid phase and on passing down the engine, rapidly cooled and solidified. It was believed that adhesion occurred only whilst the silicate was in the molten or semi-molten state.

Although neither of the deposits represent a significant corrosion problem in nickel-based materials at present, it is not known if these and similar deposited contaminants will affect the service life of thermal barrier coated components at the considerably increased surface temperatures that we envisaged in the next generations of jet engines.

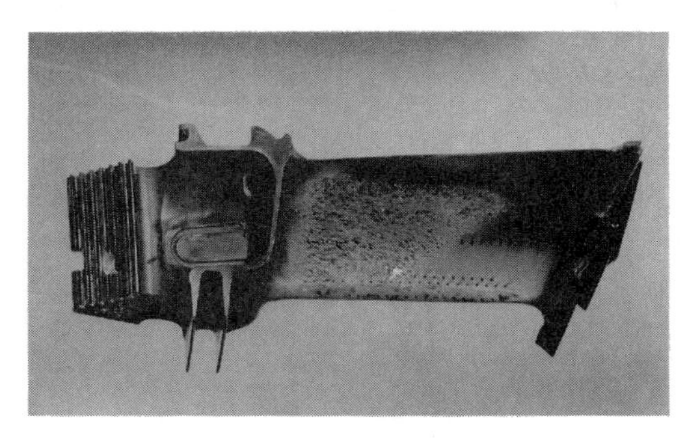

Figure 1: Silicate deposition on the aerofoil pressure surface of a civil HP turbine blade after Middle-East operation

2 EXPERIMENTAL PROCEDURE

Sand samples were obtained from several Middle-East airport locations, namely Doha, Bahrain, Dubai and Abu Dhabi Airport in the Persian Gulf, as well as a sample from Sinaiyah in western Saudi Arabia. The samples were extensively characterized before and after ageing by means of X-ray diffraction (XRD), energy dispersive X-ray analysis (EDX), and differential thermal analysis (DTA).

Free standing plasma sprayed ZrO_2-8wt%Y_2O_3 thermal barrier coated specimens with thicknesses of 200-300 μm were prepared by dissolving the nickel-based substrate and bond coat in an acid solution. The thin ceramic specimens were aged in contact with small amounts of each of the sand samples located on the outer surfaces of the coatings at temperatures of 1200-1600°C for durations varying between 1-120 hours. The aged specimens were air quenched to retain the high temperature phases.

Electron beam physical vapour deposited (EB PVD) thermal barrier coating specimens of the same composition were similarly prepared and aged with Middle-East sand. These free standing ceramic specimens were exceptionally fragile due to the columnar structure of the coatings.

3 RESULTS AND DISCUSSION

Middle-East Sand

All the sand samples exhibited significant variations in particle size and texture, clearly comprising mixtures of various minerals, as confirmed by EDX analyses of various individual particles. The average elemental compositions of the sand samples as determined by EDX analyses showed a significant presence of calcium-containing minerals in all the Persian Gulf samples, with only the Sinaiyah sand being the more usual silica-based.

According to the X-ray diffraction (XRD) analyses the minerals common to all the sand samples were quartz (α-SiO_2), calcite ($CaCO_3$), dolomite ($CaMG(CO_3)_2$), and the feldspar silicates albite ($NaAlSi_3O_8$) and microcline ($KAlSi_3O_8$), with small amounts of various other minerals. In addition to the above mentioned minerals, the Doha sand exhibited a dominant gypsum ($CaSO_4.2H_2O$) presence, and a relatively low quartz content, while the Sinaiyah sand contained very little calcium-based minerals, but mainly quartz and feldspar silicates (Table 1). DTA and EDX analyses were carried out to confirm the XRD data.

Chemical interactions between the ingested sand and the thermal barrier coatings are believed to occur only when the sand adheres to the blade surface, and therefore, only when the sand is in the molten or semi-molten state. During ageing at 1200°C for 1 hour, only the Sinaiyah sand melted and adhered to the plasma sprayed ZrO_2-Y_2O_3 substrate. Analysis indicated that the chemical composition of the Sinaiyah sand remained basically unchanged during melting, and although the sand started to glassify, quartz (α-SiO_2) and albite ($NaAlSi_3O_8$) peaks were still detected with XRD (Table 2).

XRD analysis of the unmolten sand samples identified the presence of calcium oxide (CaO), magnesium oxide (MgO), quartz (α-SiO_2) and the high temperature silicon dioxide phase, cristobalite, as well as various calcium silicates (Table 2). The calcite ($CaCO_3$), gypsum ($CaSO_4.2H_2O$) and dolomite ($CaMg(CO_3)_2$) dissociated at approximately 800°C into calcium oxide (CaO) and magnesium oxide (MgO), as confirmed by the DTA results.

At 1300°C the Bahrain, Dubai and Abu Dhabi sand samples became semi-molten. The Doha sand was, however, still unmolten after ageing. During ageing at 1400°C and above, all the sand samples, including Doha sand, became completely molten.

Table 1 Minerals identified in Middle-East sand samples by X-ray diffraction (XRD) techniques; vs = very strong, s = strong, m = medium, w = weak, vw = very weak

Mineral	Sinaiyah	Bahrain	Dubai	Abu Dhabi	Doha
Quartz α-SiO_2	vs	vs	s	vs	w
Calcite $CaCO_3$	w	w/m	vs	s	w
Aragonite $CaCO_3$	-	m	-	-	-
Dolomite $CaMg(CO_3)_2$	w	w/m	w	w	m
Gypsum $CaSO_4.2H_2O$	-	-	-	-	vs
Albite $NaAlSi_3O_8$	m	w	w	m/s	-
Microcline $KAlSi_3O_8$	m	w	w	w	-
Diopside $CaMg(SiO_3)_2$	-	w/m	-	-	-

Table 2 X-ray diffraction analyses of Middle-East sand aged at 1200°C for 1 h on thermal barrier coatings

Compound	Composition (wt.%)				
	Sinaiyah	Bahrain	Dubai	Abu Dhabi	Doha
Quartz α-SiO_2	vs	vs	s	vs	-
Cristobalite SiO_2	-	m	vs	m	m
CaO	-	-	w	w	w
MgO	-	w	vw	vw	vs
Ca_3SiO_5	-	w/m	w	w/m	m
β-Ca_2SiO_4	-	w/m	w	w/m	m
γ-Ca_2SiO_4	-	-	-	-	m
Albite $NiAlSi_3O_8$	m	-	-	-	-

The Degradation of Plasma Sprayed Thermal Barrier Coatings by Molten Middle-East Sand

Sinaiyah Sand. During ageing at 1200°C, the viscous, semi-molten Sinaiyah sand penetrated the adjacent areas of the thermal barrier coating. No apparent chemical interactions occurred between the ceramic and the silicate, and the layered structure characteristic of plasma sprayed coatings, was not affected (Figure 2a). More significantly, penetration of the silicate into the ceramic coating was observed after ageing at 1300°C.

The first microstructural evidence of chemical interactions between the molten silicate and the ceramic coating was found after ageing at 1400°C. Chemical attack of similar nature, but with an increased rate, was observed in specimens aged at 1500°C (Figure 2b). The silicate did not only penetrate the whole coating thickness, but as a result of the chemical attack, significant amounts of the ceramic were detached and floated in the silicate layer. Grain boundary attack occurred through the entire ceramic coating, leaving roundish ceramic grains in the lower parts of the coating. The upper part of the ceramic adjacent to the silicate, and in particular the floating ceramic parts in the silicate, exhibited a much finer grain structure. EDX analyses indicated that these fine phased ceramic areas were depleted in the yttria stabiliser, with only 2-3 wt.% Y_2O_3 remaining in the ZrO_2 matrix, compared with the roundish grains with approximately 8 wt.% Y_2O_3. The silicate contained approximately 5 wt.% Y_2O_3 and 13 wt.% ZrO_2. The high ratio of yttria to zirconia (5wt%:13wt%)

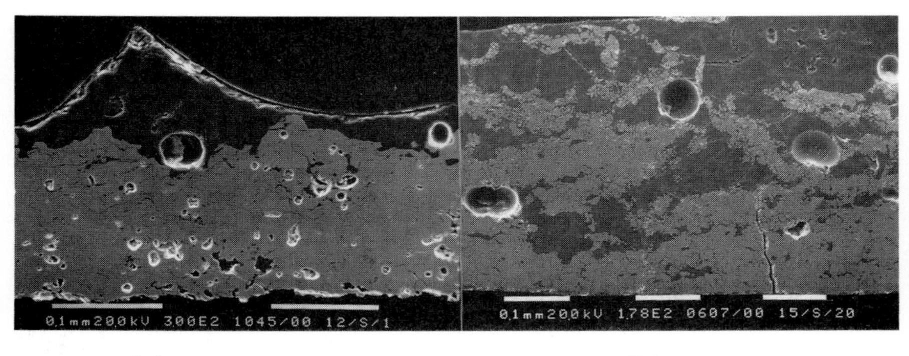

(a) (b)

Figure 2 Plasma sprayed ZrO_2-Y_2O_3 coating aged with Sinaiyah sand for, (a) 1 h at 1200°C, (b) 20 h at 1500°C

in the silicate relative to the original ceramic composition of ZrO_2-8wt%Y_2O_3, confirmed the depletion of yttria from the fine phased ceramic areas. The depletion of the yttria stabiliser initiated the disruptive tetragonal to monoclinic phase transformation during cooling, as confirmed by XRD.

The corrosion mechanism was identified during ageing of the ZrO_2-Y_2O_3 with simplified silicate glasses[4]. Yttria and zirconia were absorbed into the silicate during grainboundary attack, but zirconia in excess of the solubility limit of the slicate, was redeposited as a monoclinic, fine grain structure depleted in the stabiliser yttria.

Doha Sand. After removal of the unmolten Doha sand aged at 1300°C from the ceramic coating, microscopic examination revealed the presence of small semi-molten sand remains on the ceramic surface. The semi-molten sand did not penetrate the ceramic matrix and the microstructure was unaffected. However, EDX analysis showed substantial calcium diffusion into areas of the ceramic adjacent to the sand deposit, with a maximum of approximately 25 wt.% CaO present in the zirconia matrix. Thermal barrier coatings aged with either of the minerals calcite ($CaCO_3$) or gypsum ($CaSO_4.2H_2O$) at the same temperature, which both transformed to calcium oxide (CaO), exhibited similar levels of calcium diffusion into the ceramic matrix.

During ageing at 1400°C, the molten Doha sand formed a glassy silicate debris or slag that penetrated the ceramic coating completely. Grain boundary attack occurred through the entire coating thickness, and the roundish appearance of the ceramic grains indicated chemical attack whereby the ceramic matrix was dissolved at the grain boundaries (Figure 3a). EDX analysis of the silicate confirmed the dissolution of the ceramic matrix,

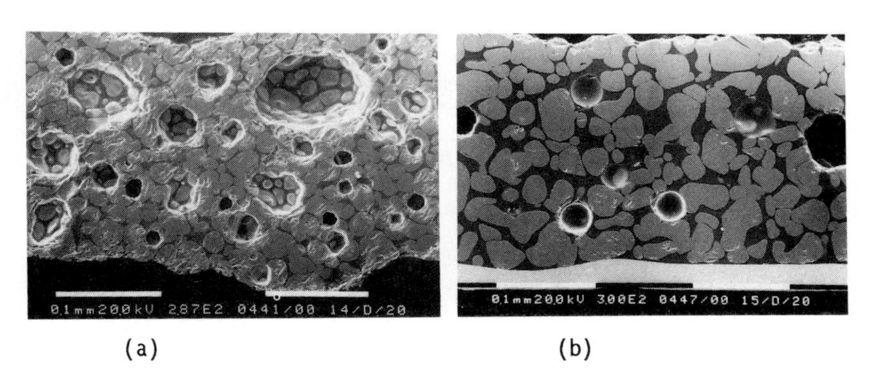

Figure 3 Plasma sprayed ZrO_2-Y_2O_3 coating aged with Doha sand for (a) 20 h at 1400°C, (b) 20 h at 1500°C

with approximately 4 wt.% Y_2O_3 and 13 wt.% ZrO_2 present in the silicate after ageing. Diffusion of calcium from the silicate into the ceramic coating occurred, resulting in approximately 4 wt.% CaO in the ZrO_2-Y_2O_3 matrix after 20 h of ageing at 1400°C, and as much as 7 wt.% CaO after ageing for 120 h at 1500°C. The large voids in the coating were probably caused by the molten sand that sealed the ceramic surface, trapping air inside the porosity of the plasma sprayed coating. This was confirmed after a plasma sprayed thermal barrier coating that was first densified during ageing without any debris for 120 h at 1500°C, contained markedly fewer voids after ageing under similar conditions in the presence of Doha sand.

After ageing for 20 h at 1500°C (Figure 3b), it appeared as if more of the ceramic matrix was dissolved since smaller roundish ceramic grains and a higher proportion of silicate to ceramic, were observed. However, EDX analysis of the silicate showed similar zirconia and yttria levels than after ageing at 1400°C. The lower viscosity of the silicate at 1500°C enabled the release of some trapped air, and fewer voids were observed.

Bahrain, Dubai and Abu Dhabi Sand. After ageing at 1300°C, penetration of the semi-molten Bahrain, Dubai and Abu Dhabi sand into the ceramic coatings, occurred. The silicate was present through the whole coating thickness, with the plasma sprayed structure transformed into small, roundish ceramic grains and a fine ceramic phase, as shown in Figure 4a for Dubai sand.

These roundish ceramic grains and fine ceramic phase became more evident after ageing at 1400°C, as shown in Figure 4b, 4c and 4d for Dubai, Bahrain and Abu Dhabi sand respectively. The fine ceramic phase was usually present in the upper half of the coating adjacent to the surface where the sand was originally deposited. The roundish ceramic grains maintained the 8 wt.% yttria percentage, while the fine phase was depleted in yttria with approximately 2-3 wt% Y_2O_3 remaining in the ceramic.

Similar microstructures were also obtained after ageing at 1500°C, and EDX analyses of the silicate compositions revealed the absorption of approximately 7 wt.% Y_2O_3 and 20 wt.% ZrO_2, compared to 3 wt.% Y_2O_3 and 13 wt.% ZrO_2 for the Doha sand aged at the same temperature and same duration. However, a lower amount of calcium diffused into the ceramic matrix after ageing with Bahrain, Dubai and Abu Dhabi sand relative to ageing with Doha sand under equivalent conditions.

During degradation by molten sand, the layered plasma sprayed microstructure transformed into large roundish ceramic grains in a silicate slag. This transformation process was studied during ageing of thermal barrier coatings with Dubai sand for short durations. The molten sand penetrated the ceramic initially through porosity and attacked the ceramic at the grain boundaries and at microcracks causing fragmentation of the ceramic structure after only 1 h of ageing (Figure 5a). The fragmentation

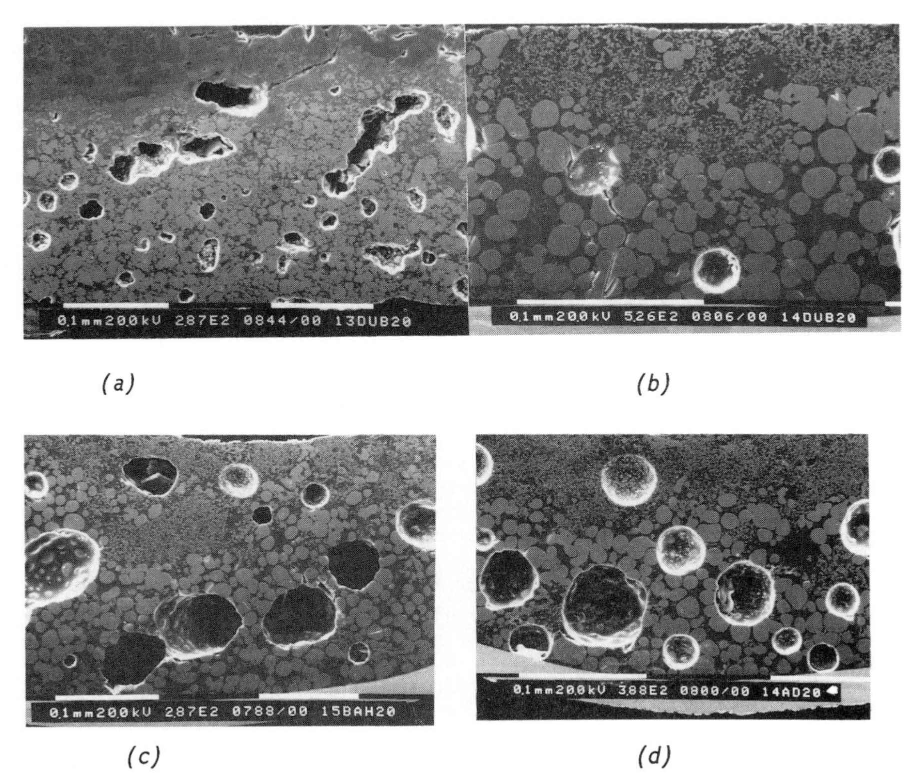

Figure 4 Plasma sprayed ZrO_2-Y_2O_3 coating aged with: (a) Dubai sand for 20 h at 1300°C, (b) 20 h at 1400°C, (c) Bahrain sand for 20 h at 1400°C, and (d) Abu Dhabi sand for 20 h at 1400°C

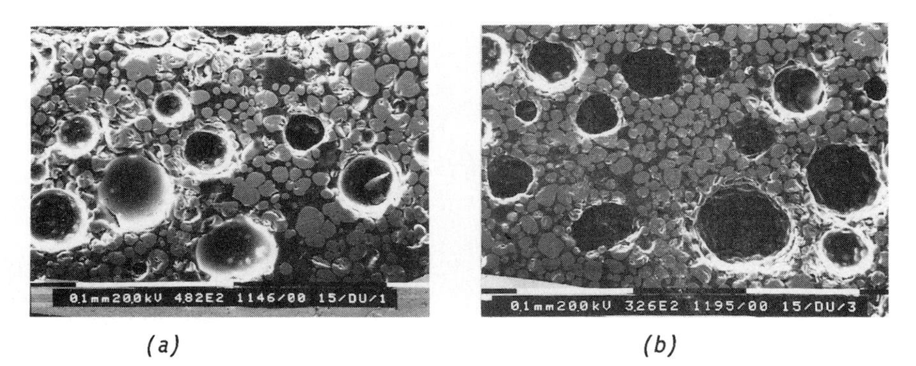

Figure 5 Plasma sprayed ZrO_2-Y_2O_3 coating aged with Dubai sand for (a) 1 h at 1500°C, (b) 3 h at 1500°C

with concurrent absorption of yttria and zirconia continued until the silicate was saturated, as shown by the small ceramic particles after 3 h of ageing (Figure 5b). Grain growth occurred during continuous ageing, resulting in the larger, roundish appearance of the ceramic grains after typical ageing cycles of 20 h (Figures 3 & 4). The grain growth continued during longer ageing durations, and very large roundish ceramic grains were observed after 120 h of ageing.

The Degradation of EB PVD Thermal Barrier Coatings by Molten Middle-East Sand

During the last few years it has become evident that the durability required for the use of thermal barrier coatings on rotating aerofoil surfaces, can only be achieved with the electron beam physical vapour deposition (EB PVD) technique.[5] The columnar growth, inherent of PVD, segments the structure with minimum compressive stress resulting in a more strain-tolerant material.

Free standing EB PVD ZrO_2-8wt%Y_2O_3 coatings were aged with Doha sand at 1500°C. After ageing for a short period of only 1 h, some molten sand had already penetrated the ceramic along the grain boundaries and transformation of the columnar structure into small, roundish ceramic grains was observed (Figure 6a). During a longer ageing cycle of 20 h, the roundish ceramic grains became more prominent, and significant grain growth occurred.

The initial stages of this transformation process were observed after ageing with a very small amount of Dubai sand for 4 h at 1500°C (Figure 6b). The molten sand penetrated the ceramic along the columnar grain boundaries, followed by horizontal grain boundary attack causing the break-up of the long columnar grains into smaller grains. The grain boundary attack and break-up of grains continued until the silicate was saturated in ceramic species and equilibrium was reached. During longer ageing periods grain growth occurred.

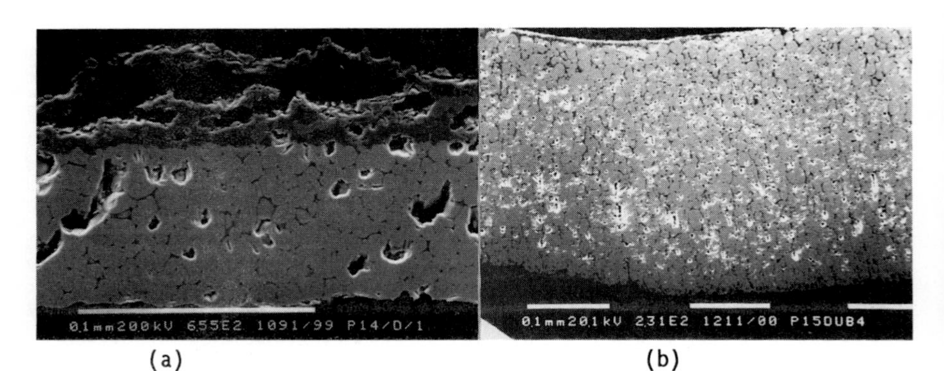

(a) (b)

Figure 6 EB PVD ZrO_2-Y_2O_3 coating after ageing with (a) Doha sand for 1 h of 1500°C, (b) Dubai sand for 4 h at 1500°C

4 CONCLUSIONS

The degradation of the ZrO_2-Y_2O_3 coatings by molten Middle-East can be grouped into three different types according to the sand composition:

a) Sand with a high quartz content, for example Sinaiyah sand, characterised by the depletion of the yttria stabiliser from the ceramic;

b) Sand consisting mainly of calcium based minerals such as gypsum, for example Doha sand, characterized by severe grain boundary attack, significant grain growth and calcium diffusion from the silicate into the ceramic matrix;

c) Sand containing both the calcium based minerals and quartz, for example Bahrain, Dubai and Abu Dhabi sand, characterised by the presence of both of these degradation mechanisms.

Although yttria was absorbed into the silicate during degradation by calcium containing sand, the calcia replaced yttria as stabilizer in the zirconia matrix, preventing the tetragonal to monoclinic phase transformation.

The molten sand penetrated the ceramic initially through porosity. During grain boundary attack fragmentation of ceramic particles, with concurrent absorption of yttria and zirconia into the silicate, occurred. After equilibrium was reached, grain growth was observed during continuous ageing. Similar degradation mechanisms were observed after ageing of EB PVD thermal barrier coatings with Middle-East sand.

ACKNOWLEDGEMENTS

Rolls Royce plc. Derby is thanked for providing the thermal barrier coatings, and the CSIR for financially supporting one of us (DdW).

REFERENCES

1. R.E. Demaray, J.W. Fairbanks and D.H. Boone, J. Am. Soc. Mech. Eng., 1982, ASME 82-GT-264, 264

2. Private Communications, Rolls Royce, 1990

3. F.C. Toriz, A.B. Thakker and S.K. Gupta, J. Am. Soc. Mech. Eng., 1988, ASME 88-GT-279, 279

4. D.J. de Wet, R. Taylor and F.H. Stott, J. Phys., in press.

5. J.W. Fairbanks and R.J. Hecht, Mat. Sci. Tech., 1987, 88, 321

2.3.3
High Temperature Corrosion Behaviour of Thermal Spray Coatings

S.T. Bluni and A.R. Marder

MATERIALS SCIENCE AND ENGINEERING, LEHIGH UNIVERSITY, BETHLEHEM, PA 18015, USA

1 INTRODUCTION

Boiler tube failures in fossil-fired power plants constitute the largest single cause of forced plant outages, which can cost up to an estimated $700,000 per day for large modern units. These failures are largely the result of corrosion in the form of oxidation and sulphidation.[1-4] In efforts to address this costly corrosion problem, utilities have turned to the use of various high-temperature coatings for boiler tube protection. Perhaps the most suitable of these coatings types is thermal spray, which can be applied to boiler tubes in-situ. Experience with thermal spray coatings is limited among utilities, but many companies have been successful with their use. However, a systematic coating selection and application procedure has not been determined, resulting in a costly trial-and-error basis for coating use. In addition, the critical coating properties which determine the success or failure of a coating in the boiler environment have not been identified.

The objective of this study is to investigate the structure-property relationships of various thermal spray coatings as related to high-temperature oxidation and sulphidation behaviour. Through the identification of the critical microstructural characteristics which govern the severity of coating attack when placed in aggressive environments, quality assurance guidelines can be established so that coating performance may be anticipated.

2 EXPERIMENTAL METHOD

Three coating/substrate systems were used for analysis based on their commercial availability and promotion for boiler tube application. The coating compositions as quoted by the spray vendors, and substrate compositions as determined by wet chemical analysis, are listed in Table 1.

For oxidation testing, five samples of each coating were simultaneously oxidized in laboratory air at 600^0C. One sample of each coating was removed for analysis at 50, 100, 250, 500 and 1000 hours. Each specimen consisted of a coated low-alloy tube steel section which had an approximate size of 25mm in length, 12.5mm in width, and 12.5mm in thickness. The sulphidation test parameters were the same as those for the oxidation test. Samples were placed in a tube furnace equipped with a 63.5mm diameter alumina tube. The sulphur dioxide flow rate was kept constant at approximately 0.25ml/sec. Nitrogen was used to purge the system before and after samples were removed.

Table 1 Substrate and coating composition (in weight percent) for all coating systems analyzed

Alloying Element	Coating System 1		Coating System 2		Coating System 3	
	Substrate	**Coating**	**Substrate**	**Coating**	**Substrate**	**Coating**
Cr	1.21	45	0.10	27	0.45	26.5
C	0.11	--	0.22	--	0.23	--
Mo	0.50	--	0.02	2	0.01	--
Ni	--	51	--	--	--	62
Fe	bal	--	bal	65	bal	--
Al	--	--	--	6	--	7
Co	--	--	--	--	--	3.5
Ti	--	4	--	--	--	--
Y_2O_3	--	--	--	--	--	1

Both the as-sprayed and laboratory-tested coated samples were microstructurally characterized with the use of light optical and scanning electron microscopy. Semi- and fully-automated image analysis techniques were employed to obtain quantitative microstructural coating information.

3 RESULTS

As-Sprayed Microstructural Characterization

The as-sprayed microstructures of coatings 1, 2, and 3 can be seen in Figure 1(a)-(c) respectively. The size, shape, and distribution of oxides and voids within each coating can be seen by examination of these micrographs. Measurements of coating thickness, as well as the oxide and porosity contents for each coating, are listed in Table 2.

Oxidation Test Results

The typical structure of each coating after 1000 hours exposure can be seen in Figure 2(a)-(c). Several microstructural changes result from exposure time. These include: i) the formation of oxides in the form of outgrowths at coating surfaces, ii) an increase in intercoating oxide contents, and iii) the formation of an oxide scale at coating/substrate interfaces (note arrows in Figures). The change in each one of these with oxidation exposure time can be seen in Figure 3(a)-(c). The data shown in these graphs represent average values.

Table 2 As-sprayed microstructural characterization results for the coatings examined

COATING	THICKNESS (microns)	POROSITY (area %)	OXIDE (area %)
1	543 ± 60	0.7 ± 1.0	26.0 ± 0.4
2	605 ± 67	5.5 ± 0.8	11.0 ± 0.7
3	993 ± 16	0.1 ± 0.1	30.5 ± 2.2

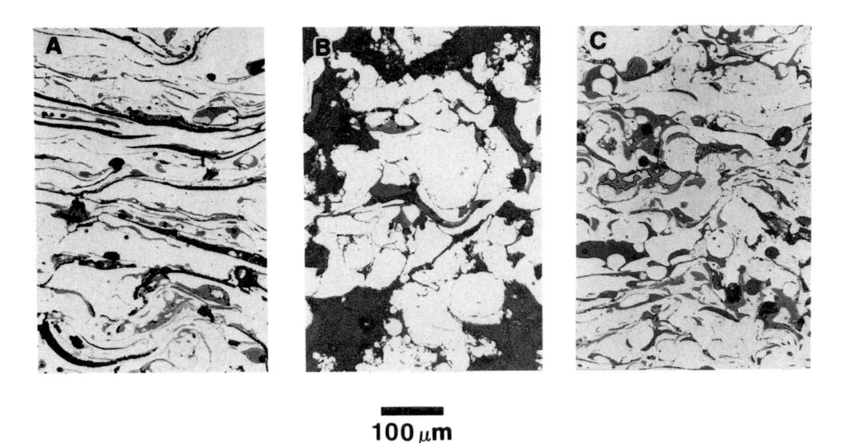

Figure 1 As-sprayed microstructure of (a) coating 1, (b) coating 2, and (c) coating 3

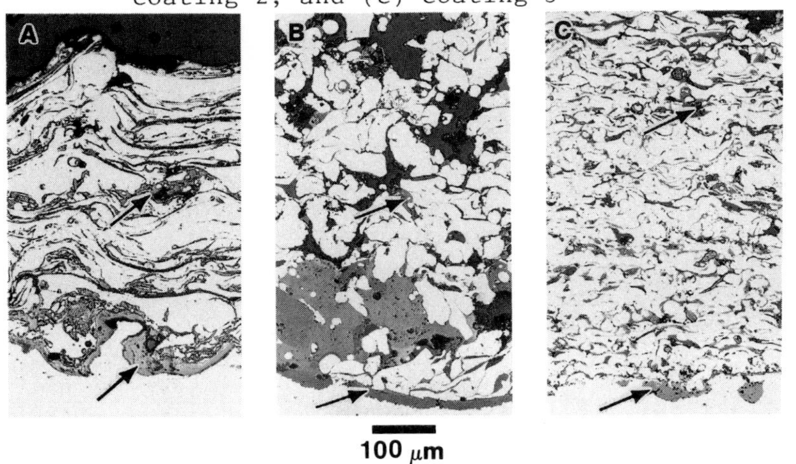

Figure 2 Microstructure of (a) coating 1,(b) coating 2, and (c) coating 3 after 1000 hours of oxidation exposure time

Sulphidation Test Results

The microstructures of coatings 1,2, and 3 after exposure to the sulphidation environment can be seen in Figure 4(a)-(c), respectively. As in oxidation testing, the important occurrences here include the formation of corrosion outgrowths at the coating surfaces, an increase in the coating oxide content, and the formation of a corrosion scale at the substrate/coating interface (arrows

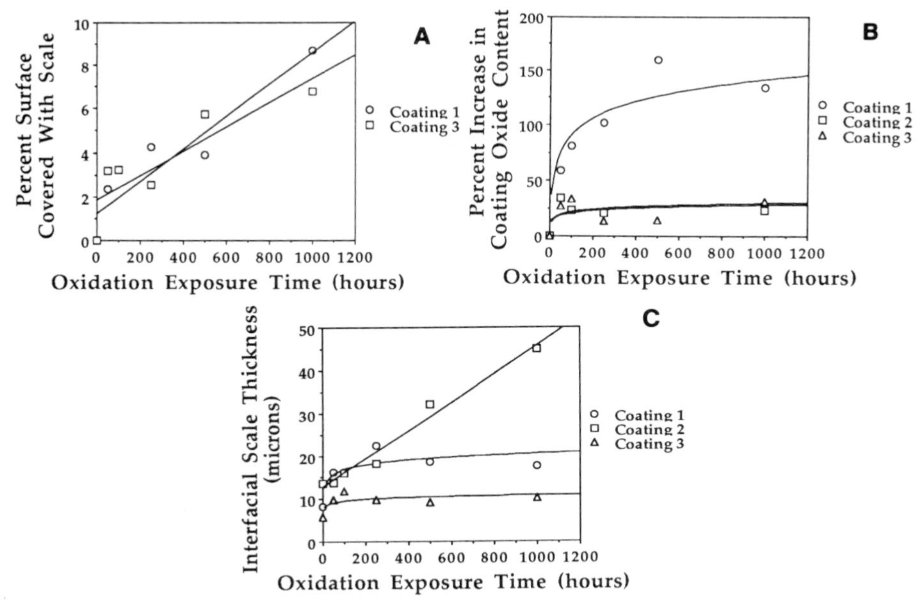

Figure 3 Oxidation test results. Plotted as functions of exposure time for each coating include: (a) the percent of the coating surface covered with outgrowths or scale, (b) the percent increase in the coating oxide content compared to the as-sprayed condition, and (c) the thickness of the corrosion scale at the coating/substrate interface

in Figures). The change in each of these phenomena with sulphidation exposure can be seen in Figure 5(a)-(c). It should be noted that, here, "oxide" denotes oxides and/or sulphides which may have formed from the corrosion process.

4 DISCUSSION

Surface Corrosion

Because of large data scatter, it is difficult to obtain any kinetic information from outgrowth size.

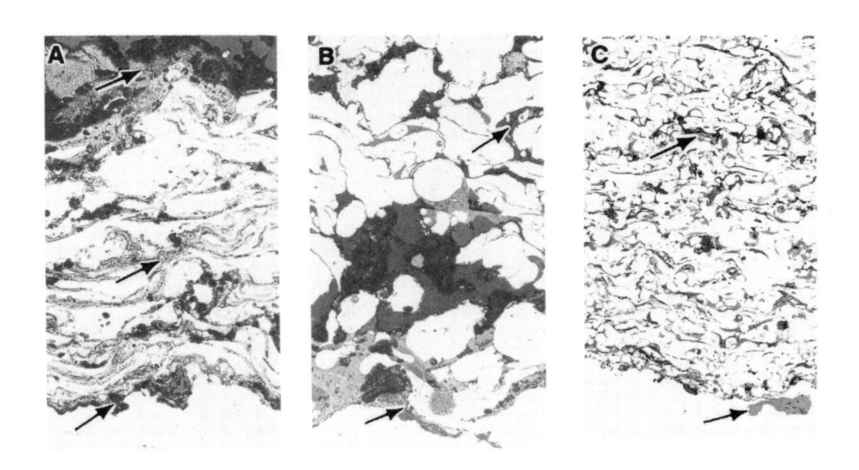

Figure 4 Microstructure of (a) coating 1, (b) coating 2, and (c) coating 3 after 1000 hours of sulphidation exposure time

However, a measure of surface reactivity can be obtained from Figures 3(a) and 5(a), which illustrate the percent coating surface covered with corrosion after exposure intervals in oxidation and sulphidation environments, respectively. Because of the relatively slow reaction kinetics for the oxidation of coatings 1 and 3, and the subsequent large percentage of surface area available for corrosion product nucleation, Figure 3(a) shows a linear relationship for all exposure intervals. A linear relationship also exists for sulphidizing conditions (Figure 5(a)), but here, 100% of the coating surface is covered with scale after 100 hours of exposure for coating 3 and after 1000 hours for coating 1. The rapid sulphidation kinetics is explained by the reactions between nickel and sulphur dioxide, as discussed by other investigators.[5-10]

Kinetics of Corrosion Scale Formation at the Coating/ Substrate Interface

The thickness of the coating/substrate interfacial scale as a function of exposure time has been plotted in Figures 3(c) and 5(c) for oxidation and sulphidation

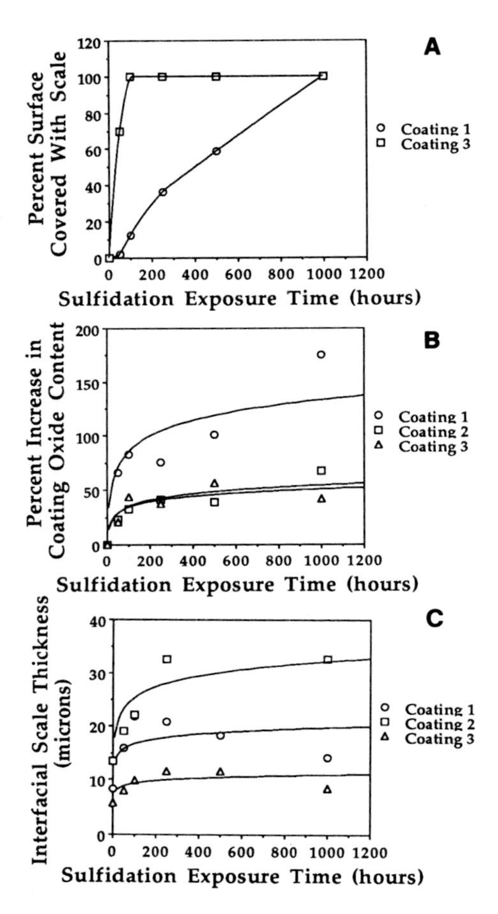

Figure 5 Sulphidation test results. Plotted as functions of exposure time for each coating include: (a) the percent of the coating surface covered with outgrowths or scale, (b) the percentage increase in the coating oxide content compared to the as-sprayed condition, and (c) the thickness of the corrosion scale at the coating/substrate interface

environments, respectively. The corresponding corrosion rates have been determined from these plots. Table 3 summarizes the type of oxidation which has occurred for each coating and the environment as well as the rate constant for each. As shown in this table, the corrosion rates were in most cases logarithmic in nature, corresponding to a limiting oxide layer thickness where further oxidation becomes negligible. This behaviour is attributed to the formation of protective chromium oxide layers at coating splat boundaries, which clog fast diffusion paths and limit subsequent substrate attack. The only exception to this type of behaviour was found for the coating 2 system in oxidizing conditions, which best resembles linear rate oxidation kinetics. In this case, the rate of alloy corrosion is unaffected by oxide formation and proceeds at a constant rate.

Table 3 Coated substrate corrosion rate type and corresponding rate constants for all coatings and both test environments

Coating/Environment	Oxidation Type	Rate Constant
1 / oxidation	logarithmic	3.75
2 / oxidation	linear	0.033
3 / oxidation	logarithmic	1.34
1 / sulfidation	logarithmic	3.03
2 / sulfidation	logarithmic	6.81
3 / sulfidation	logarithmic	1.57

In addition to the thickness of the corrosion scale at the substrate/coating interfaces, the thickness of the corrosion scale on the uncoated substrate regions was measured. The corresponding oxidation rate types and constants are shown in Table 4. As can be seen from this table, all substrates oxidized in a linear, or

Table 4 Uncoated substrate corrosion rate type and corresponding rate constants for all substrates and both test environments

Substrate/Environment	Oxidation Type	Rate Constant
1 / oxidation	linear	0.081
2 / oxidation	linear	0.246
3 / oxidation	linear	0.691
1 / sulfidation	logarithmic	48.3
2 / sulfidation	logarithmic	57.2
3 / sulfidation	parabolic	4.21

catastrophic, rate in oxidizing environments with the fastest attack occurring for substrate 3 and slowest for substrate 1. Conversely, substrate attack was not as severe in sulphidizing conditions, where substrates exhibited either logarithmic or parabolic corrosion rates.

A comparison of oxidation rates between coated and uncoated substrates can be seen by inspection of Tables 3 and 4. In this respect, coatings 1 and 3 effectively protected their substrates, as linear substrate oxidation rates for uncoated substrates were reduced to logarithmic rates for coated ones. On the other hand, the oxidation rate for substrate 2 remained linear even when coated. Although the application of coating 2 resulted in an order of magnitude difference in linear oxidation kinetics, the failure of the coating to limit substrate corrosion such that no further oxidation occurred after a certain exposure time makes its performance inferior to the protection offered by coatings 1 and 3. The differences in coating performance are attributed to coating structure. For example, coating 3 provided the best substrate protection of all coatings in both test environments; in oxidizing conditions, the coating 3 substrate was reduced from the highest linear rate to the lowest logarithmic rate, while for sulphidizing conditions, it was reduced from the only parabolic rate to the lowest logarithmic rate. Conversely, coating 2 provided the least substrate protection resulting in the highest substrate corrosion kinetics for both test environments. Since coating 3 is the most dense and thick coating of those studied, and coating 2 is the most porous by an order of magnitude, these results are expected. Although coating 2 does contain chromium and is expected to form internal chromium layers, the porous nature of this coating prohibits the oxide from becoming protective.

Interfacial Scale Thickness as a Function of Coating Microstructure

The thickness of the coating/substrate interfacial scale resulting from test procedures was used to determine the relationship between coating efficiency and microstructure. A parameter, "free path to substrate" (FPS), was developed and used as a measure of coating microstructure. This parameter denotes the planar cross-sectional distance from coating surface to substrate via splat boundaries, voids, and oxides. In this way, FPS is directly proportional to coating thickness and inversely proportional to porosity and oxide content. The relationship between the IS thickness resulting from the corrosion process and the minimum FPS, for all coatings and both test environments, is shown in Figure 6. In the function shown on the ordinate of this plot, [$(IST-IST_0)$ / relative UST], the terms are defined as follows:

IST = Interfacial scale thickness [microns].

IST_0 = IST in as-sprayed condition [microns].This term is necessary so that the contribution to IST from spraying, and not the corrosion process, is subtracted.

UST = Thickness of corrosion scale found on uncoatedsubstrate regions. "Relative UST" refers to the thickest UST among substrates for the particular time interval of interest. This term is necessary to normalize the IST measurement with respect to various substrate reactivities.

Figure 6 shows that IST due to corrosion is limited at high FPS's, indicating that substrate attack is therefore minimal for coatings which are dense and thick.

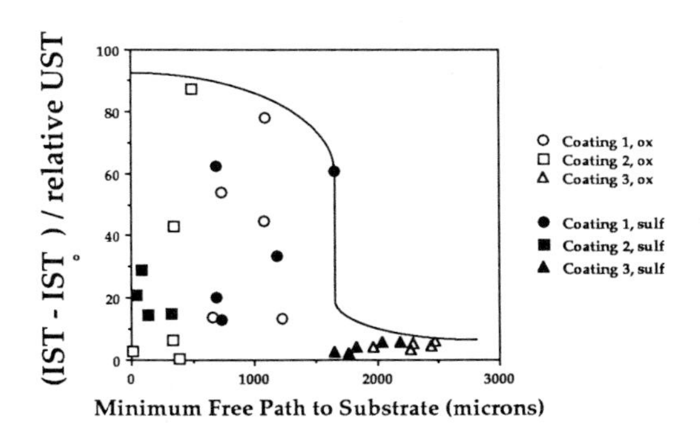

Figure 6 Thickness of the corrosion scale at the substrate/coating interface as a function of FPS for all coatings and environments investigated

5 CONCLUSIONS

1. The nickel-containing coatings were much more reactive in high-temperature sulphidizing conditions

as compared to oxidizing conditions. For example, a continuous surface scale formed on coating 3 after only 100 hours of exposure to SO_2, while only about 7% of its surface was covered with corrosion after 1000 hours in air.

2. The thickness of the scale which formed at the coating/substrate interfaces as a result of the corrosion process was used as a measure of coating efficiency. The thickness of this scale was found to be dependent on coating microstructure, with a limited interfacial scale thickness for thick and dense coatings (high FPS's).

3. Substrate corrosion rates were found to be substantially less for coated substrates as opposed to uncoated ones. Dense coatings were able to reduce substrate oxidation kinetics from linear to logarithmic by the formation of protective oxides at splat boundaries, voids, and other fast diffusion paths. Porous coatings were unable to form such protective oxides, and substrate oxidation kinetics remained linear.

ACKNOWLEDGEMENTS

This work was funded by the Pennsylvania Electric Research Council (PERC), Potomac Electric Power Company (PEPCO), Public Service Electric and Gas (PSE&G), Ohio Edison and Virginia Power. The authors would also like to thank Mr Arlan Benscoter of Lehigh University for numerous metallographic consultations.

REFERENCES

1. D.B. Meadowcraft, Matls. Sci. Eng., 1987, 88, 313.
2. J. Cocubinski, "Failures and Inspections of Fossil-Fired Boiler Tubes," [Proc. conf.], EPRI CS-3272, Palo Alto, California, 1983.
3. J.H. Pohl, M.P. Heap and A.K. Mehta, ibid.
4. J. Stringer, "High Temperature Corrosion in Energy Systems," [Proc. conf.] M.F. Rothman ed., AIME, New York, 1985, p.3.
5. M. Seiersten and P. Kofstad, Corr. Sci., 1982, 22, 487.
6. P. Kofstad and G. Akesson, Oxid. Met., 1978, 12, 503.

7. F. Gesmondo, C. de Asmundis and P. Nanni, Oxid. Met., 1983, 20, 102.
8. A. Anderson and P. Kofstad, Corr. Sci., 1984, 24, 731.
9. W.L. Worrell and B.K. Rao, "High Temperature Corrosion," [Proc. conf.], R.A. Rapp ed., National Association of Corrosion Engineers, Texas, 1981.
10. K. Natesan and D.J. Baxter, "Corrosion-Erosion-Wear of Materials at Elevated Temperatures," [Proc. conf.], A.V. Levy ed., National Association of Corrosion Engineers, California, 1986.

Section 2.4 Machining

2.4.1
The Effects of Temperature Rise on the Rheology of Carrier Media Used in Abrasive Flow Machining

J.B. Hull, A.R. Jones, A.R.W. Heppel, A.J. Fletcher,[1] and S. Trengove[2]

UNIVERSITY OF BRADFORD, BRADFORD, UK

[1] SHEFFIELD HALLAM UNIVERSITY, SHEFFIELD S1 1WB, UK

[2] EXTRUDE HONE LTD., MILTON KEYNES, UK

1 INTRODUCTION

Abrasive Flow machining (AFM) is a novel surface finishing technique for the final stage machining of a wide range of engineering components. The process has been employed to deburr components after NC or CNC machining, to polish machined component sections, and to modify machined radii. Significant improvement in surface finish is often combined with other beneficial effects such as the induction of residual compressive stresses and the removal of surface flaws and thermal recast layers.

The process utilises a viscoelastic borosiloxane polymer impregnated with different abrasive media. Depending upon the requirement, particle sizes ranging from 0.25 µm (crushed diamond) to 1.5 mm (coarse silica carbide) are used in polymer/grit mixes containing up to 66% grit content by volume. The abrasive media mixes are forced through component orifices or over component surfaces, under high pressure, using special equipment (see later), in order to achieve the required machining action.

The main aim of current research is to attain greater understanding of the fluid and thermal flow characteristics of the standard grades of "base" media, in order to establish a fundamental basis for control and optimisation of AFM processes.

Essentially, three main grades of polyborosiloxane are in common use, i.e. low, medium and high viscosity formulations coloured red, yellow and blue respectively, for identification. Intermediate viscosity formulations are sometimes required, and this is achieved by mixing the requisite blends of two of the three main grades.

This paper reports on work carried out to investigate the rheological characteristics of low and high viscosity grades of base media, using a computer aided twin capillary Rosand rheometer.

The main parameters under investigation were:-

1) the behaviour of the media at varying strain rates ($1 \leq \dot{\gamma} \leq 10^4\ s^{-1}$).
2) the behaviour of the media at constant strain rate, both at low ($\dot{\gamma} \approx 10\ s^{-1}$) and high ($\dot{\gamma} \approx 10^4\ s^{-1}$) strain rate settings.
3) the behaviour of the media at elevated temperatures within the range 30°-$70^{\circ}C$.

The last parameter is considered to be of significant interest because there is an inherent increase in temperature always apparent during machining (1). The results of an initial investigation into the flow characteristics of the medium viscosity grade, at room temperature, using a single barrel Davenport rheometer have been reported elsewhere (2).

2 DESCRIPTION OF PROCESS

Process Development

The Extrude Hone abrasive flow machining process was initially developed in the mid 1960's in order to find a more effective method of deburring hydraulic control blocks, which at the time were being deburred by hand. The blocks were used at the centre of the hydraulic control system for the radar guided target acquisition system of Phantom jets, and contained 83 intersecting holes in a block no bigger than 150 cm^3(3).

By 1968, the process had been successfully developed for application to the hydraulic control blocks by the Extrude Hone Corporation. Subsequently, the process was rapidly diversified into a number of additional applications such as aerospace components and die and mould polishing operations.

3 PROCESS FUNDAMENTALS

In the basic process, two vertically opposed cylinders extrude abrasive media back and forth through passages formed by the workpiece and tooling. Abrasive action occurs wherever the media enters and passes through the most restrictive passages. A typical machine used for AFM is shown schematically in Figure 1.

The major elements of the process include:

- The tooling which confines and directs the media flow to the appropriate areas;

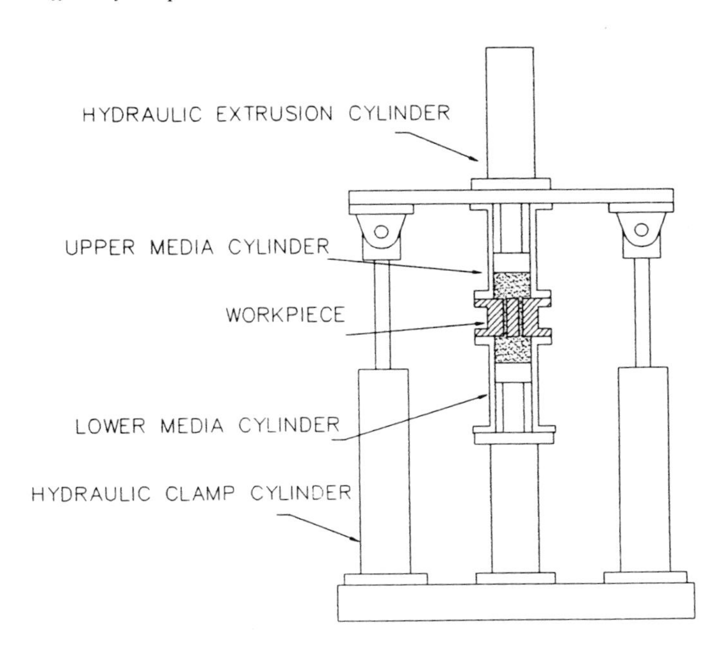

Figure 1 Schematic of typical AFM extrusion press

Shear Rate /s	Time s	Pl MPa	Ps MPa	Shear Stress KPa	Shear Viscosity Pa.s	n
150	23.70	0.84	0.12	13.10	87.34	1.69
300	39.50	5.10	0.62	79.70	265.60	1.29
450	47.80	8.50	0.89	132.80	295.00	1.05
600	56.00	8.34	1.14	130.30	217.20	0.88
750	63.30	9.11	1.43	142.30	189.80	0.75
900	71.60	9.42	1.73	147.20	163.50	0.64
1110	79.90	9.98	2.14	155.90	140.40	0.52
1320	88.10	9.90	2.53	154.70	117.20	0.42
1590	95.30	10.16	2.94	158.70	99.80	0.31
1860	104.00	11.85	3.31	185.20	99.57	0.21
2100	108.00	12.39	3.50	193.50	92.16	0.14
2400	116.00	13.44	3.76	209.90	87.47	0.06
2700	119.00	14.22	3.92	222.20	82.30	0.01
3000	127.00	14.50	3.98	226.60	75.54	0.01
3600	131.00	15.57	4.00	243.20	67.56	0.01
4500	135.00	16.75	3.99	258.90	57.53	0.01

Table 1 Rheological data from test results shown in Figure 2(a)

- The machine which controls the media extrusion pressure, flow volume and, if desired, flow rate;

- The media which determines the pattern and aggressiveness of the abrasive action that occurs.

By selectively permitting and blocking flow into, or out of, workpiece passages, tooling can be designed to provide media flow paths through the workpiece that restrict flow at the areas where deburring, radiusing and surface improvement are desired. Frequently, multiple passages or parts are processed simultaneously.

4 EXPERIMENTAL PROCEDURE

Throughout this work, extensive use was made of a Rosand rheometer. The twin capillary extrusion system is computer aided for the purposes of control and analysis, and supports the automatic application of both Bagley (4) and Rabinowitsch (5) corrections to account for die entrance pressure effects, and non parabolic flow, respectively. Although the Bagley correction was always applied to the results, the Rabinowitsch correction was considered non-applicable to borosiloxane based polymers, because it could not be assumed that the media under test were incompressible and that flow would be steady and time independent. The maximum error in viscosity resulting from ignoring the Rabinowitsch approach is - 15% (6). In the Rosand system, each barrel has three heaters.These top, centre and bottom (die) heaters are provided with corresponding thermocouples to facilitate temperature control and monitoring. Because the nature of extrusion rheometry necessitates the loading of polymer from one end of a barrel prior to testing, air entrapment was a serious problem. Although "sufficient" settling time was allowed for all tests (up to 1 hr), a large number of experimental results were discarded because of this problem. It was readily apparent, however, that raising the temperature by a small amount, (5°-8°C above ambient), was beneficial in promoting release of air trapped within the media under test. Hence, most of the work was carried out at elevated temperatures ($\geq 30^{\circ}$C.)

5 RHEOLOGICAL CHARACTERISTICS OF BOROSILOXANE MEDIA

The low viscosity medium exhibited clear evidence of the stick slip phenomenum first reported by Rhoades (7), as can be seen in Figure 2(a).
At relatively low strain rates ($\gamma \leq 5 \times 10^{2}$ s^{-1}) the amplitude of oscillation was significant ($\leq$ 2 MPa pressure fluctuations on the long die), whilst at higher strain rates ($\gamma \geq 2 \times 10^{3}$ s^{-1}) the oscillations decreased to an almost negligible level.

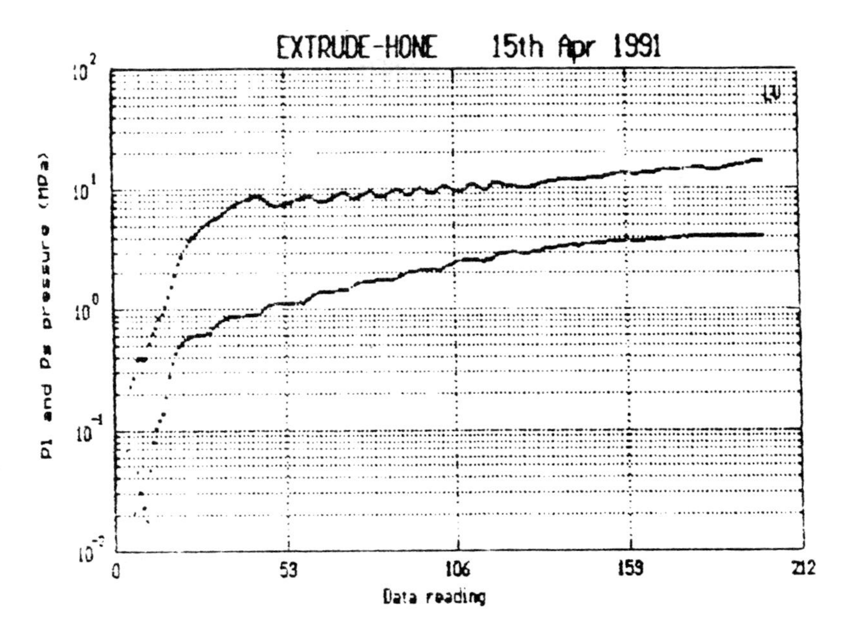

Figure 2(a) Rosand long (Pl and Ps) pressure relationship for low viscosity media at 30^{0}C

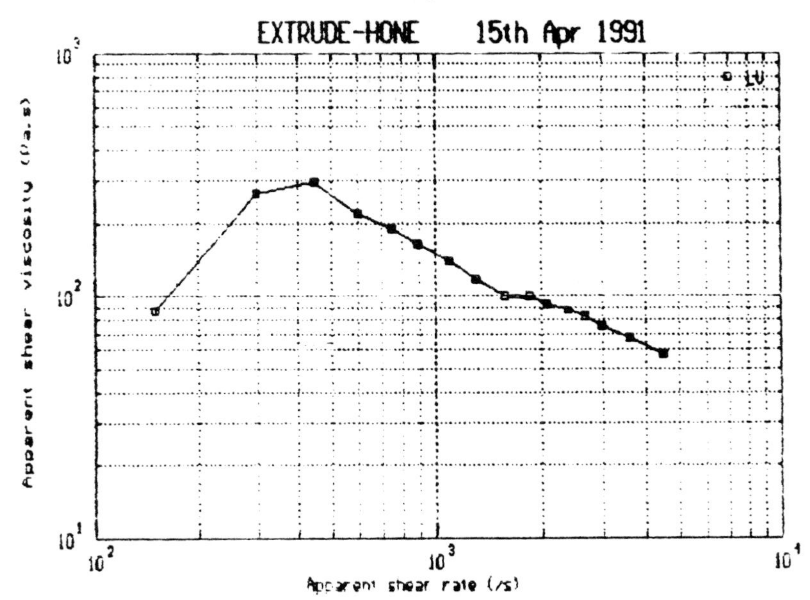

Figure 2(b) Apparent shear viscosity/shear strain rate profile for test data shown in Figure 2(a)

In addition, the medium exhibits complex rheological behaviour. Using the simple rheological relationships:-

$$\tau = k\dot{\gamma}^{n} \tag{1}$$

$$\eta_{A} = \tau\dot{\gamma}^{-1} \tag{2}$$

$$\eta_{A} = k\dot{\gamma}^{n-1} \tag{3}$$

where :- τ = shear stress
$\dot{\gamma}$ = shear rate
η_A = apparent shear viscosity
k and n are polymeric power law constants

it can be seen that at low strain rates ($\dot{\gamma} \leq 5 \times 10^{-2} s^{-1}$) the medium is apparently dilatent ($n > 1$), whilst at high strain rates($\dot{\gamma} \geq 6 \times 10^{2}\ s^{-1}$) , the material exhibits pseudoplastic behaviour ($n < 1$), and at some intermediate strain rate the material is Newtonian ($n = 1$), as shown in Table 1 and Figure 2(b).

6 EFFECTS OF TEMPERATURE IN MEDIA RHEOLOGY

The effects of temperature change on the viscosity/shear rate profiles for base low viscosity borosiloxane, are shown in Figure 3a. Evidence suggests that at constant shear rate, for shear rates within the decade range 10^{3}-$10^{4}s^{-1}$, the variation of viscosity with temperature can be described by a relationship of the type:-

$$\eta_{A} = \eta^{*}_{A}\exp[-b_{T}(T-T^{*})]$$

where T^{*} and η^{*}_{A} are the reference temperature and reference viscosity, respectively. For temperatures in the range 30°C-70°C, $b_T \approx 0.018\ {}^{0}K^{-1}$.

At shear rates $< 10^{3}\ s^{-1}$ the viscosity/temperature relationship appears to be more complex. The effects of shear history as shown previously by Hull et al (8) are more pronounced.

For the purposes of process control, an understanding of the variation of viscosity with temperature is important.

Utilisation of extrusion-honing equipment, in the commercial field, has resulted in the recognition of a significant change in media behaviour with temperature. When the media becomes "too hot to handle" (arbitarily > 60°C), there is a tendency for grit particles to "sink" under gravity to the bottom of the extrusion chambers. In addition, the media will sometimes undergo a permanent change in physical properties which necessitates recharging with new media before the processing operation can continue. It is readily apparent that this behavioural

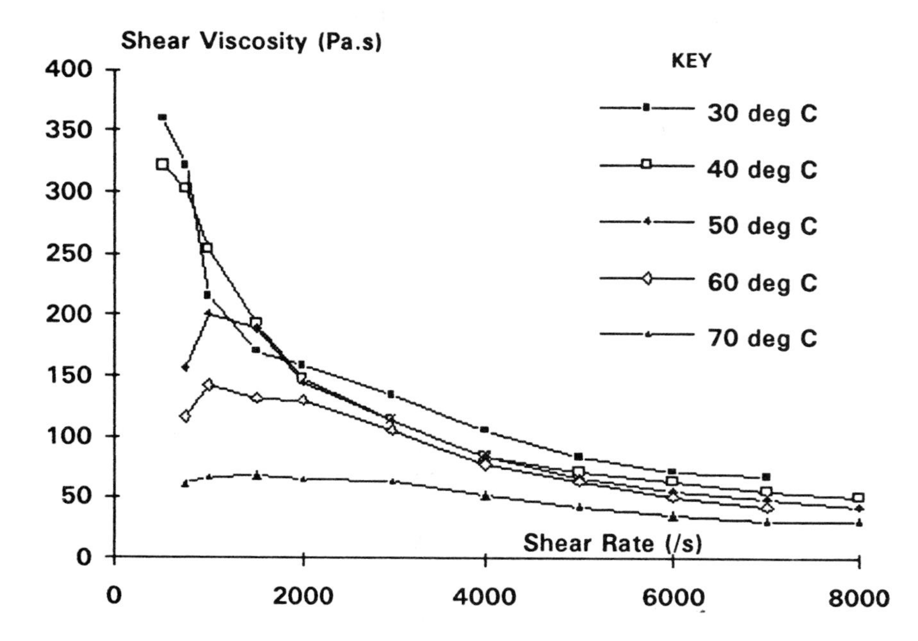

Figure 3(a) Effect of temperature on shear viscosity/shear rate

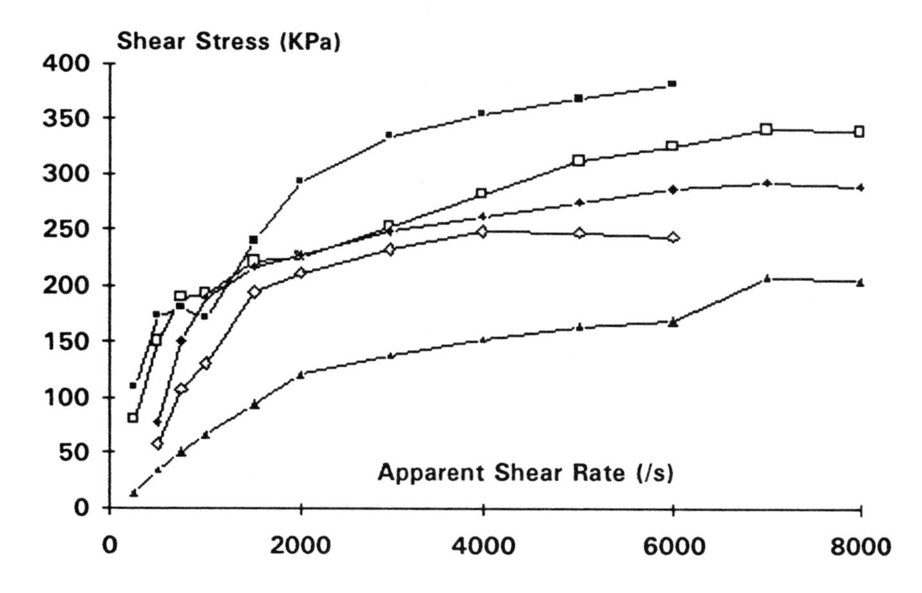

Figure 3(b) Effect of temperature change on shear stress/shear rate

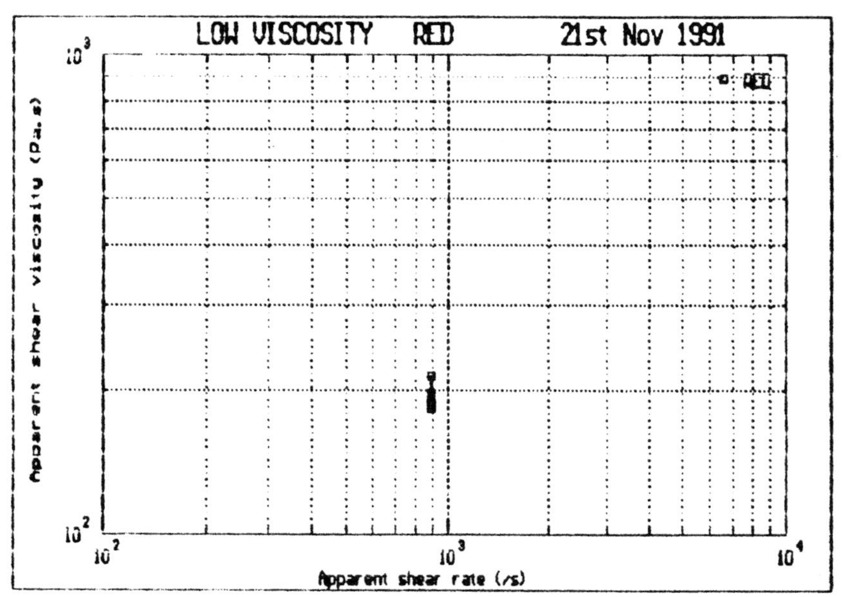

Figure 4(a) Shear viscosity at 30^{0}C and constant strain rate of 900/s

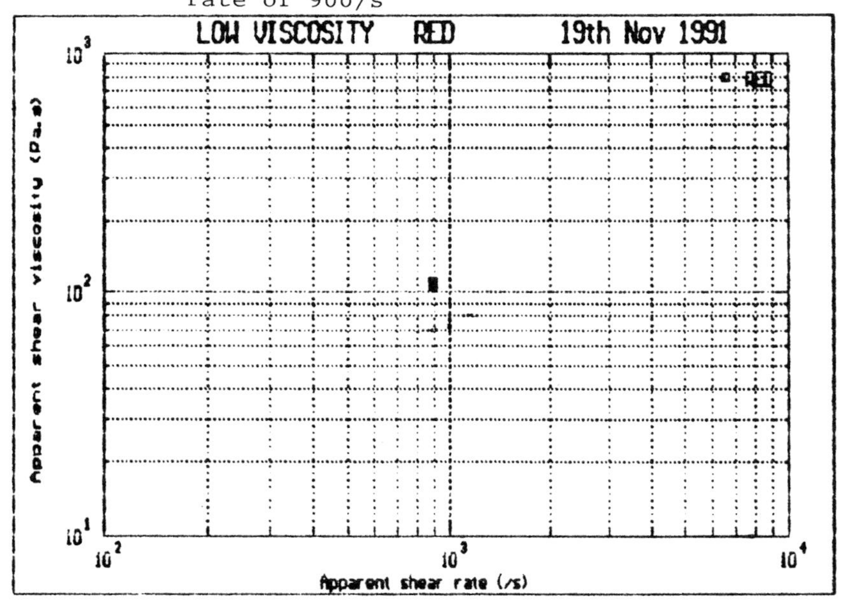

Figure 4(b) Shear viscosity at 70^{0}C and constant shear rat of 900/s

change is viscosity based. Hence, a knowledge of the critical viscosity at which media breakdown occurs is a prime necessity.

Results obtained, so far, have indicated that when the borosiloxane viscosity falls to below about 30 Pa.s abrasive grit is less likely to remain in suspension within the media.

Further evidence of the complex rheological behaviour of borosiloxane polymers at low strain rates is shown in Figure 3b. Rapid changes of shear stress with shear strain rate are apparent, with particularly significant changes at low temperatures.

An additional parameter which contributes to the pronounced rheological behaviour of borosiloxane polymers is media compressibility (2,8). In the initial stages of any rheological test, the medium under investigation exhibits an increase in viscosity. This is clearly observed at constant strain rates as shown in Figure 4a. This time dependent behaviour is most apparent at low temperatures, ($\leq 30^{\circ}$C). At higher temperatures, the behaviour is less pronounced as shown in Figure 4b.

7 CONCLUSION

1. Temperature change has a marked effect on the time dependent rheotropic behaviour of the borosiloxane based media used in abrasive flow machining.

2. In order to effect optimisation and control of the Extrude Hone process, it is necessary to predict changes in viscosity with temperature, because at some value of viscosity ($\eta_A \approx 30$ Pa.s), grit separation under gravity takes place. Hence, it is necessary to maintain the process operating temperature below T_{crit} ($\eta_A \approx 30$ Pa.s)

REFERENCES

1. S.A. Trengove ,Ph.D. Thesis, Sheffield City Polytechnic, to be published 1992.

2. A.J. Fletcher, J.B. Hull, J. Mackie and S.A. Trengove, Proc. of Int. Conf. on Surface Engineering,'Elseviers', Ed. S.A. Megiud. Toronto 1990, pp 592-609.

3. L.J. Rhoades, True Grit-The Extrude Hone Story, Pittsburgh High Technology, NOV./DEC., 1988.

4. E.B. Bagley, J. of App. Phys., 1957, 28, 624.

5. B. Rabinovitsch, as described by J.A. Brydson, Flow Properties of Polymer Melts, Godwin, 1981, Sections 2.2.2-2.2.6.

6. F.N. Cogswell, Polymer Melt Rheology, Godwin, 1981, p.135.

7. L.J. Rhoades, Flow Behaviour of Industrial Borosiloxane, NSF Phase 1 Final Report. August 1985.

8. J.B. Hull, A.J. Fletcher, D. O'Sullivan, S.A. Trengove and J. Mackie, IMF7 Limerick, September 1991.

2.4.2
Development and Applications of PVD Processes to Cutting Tools

R. Chiara, G. D'Errico, and F. Rabezzana[1]

INSTITUTO LAVORAZIONE METALLI, CNR – ORBASSANO-TO, ITALY

[1] METEC, TORINO, ITALY

1 INTRODUCTION

The evolution of PVD processes has led to the possibility of depositing new coating layers with different characteristics from those of titanium nitride. In particular, among all possible coating materials for tribological use that can be obtained by PVD, researchers are today focused on the following coating types.

(a) single-layer binary coatings: CrN, ZrN, HfN, MoN;

(b) ternary single-layer coatings: TiCN, TiAlN, TiHfN, TiNbN, TiZrN, CrAlN;

(c) multi-layer coatings: TiN + TiAlN, TiN + HfN, TiN + TiZrN + ZrN;

(d) multiphase layers i.e. coatings having differing characteristics from the surface to the interface with the substrate.

As to the industrial applications of these studies and research, today, in addition to the standard coating by TiN, the following materials are also available: TiCN, TiAlN and CrN. To illustrate the properties of some of these "new coatings", we are reporting some experimental data from research that C.N.R. is carrying out along with other European research centres. During the first stage of this research, TiAlN and TiHfN single-layer ternary coatings have been deposited and optimized.

2 DEVELOPMENT AND OPTIMIZATION OF NEW WEAR RESISTANT PVD COATINGS ON DIFFERENT TYPES OF HSS TOOLS

The coatings were produced using PVD Arc Evaporation Reactive Ion Plating initially on samples and tools made with different HSS materials (M2, M35, ASP30, etc.). The coating quality was examined using laboratory quality control methods and the coatings produced were optimized with different machinability tests on different types of HSS tools.

Formation and Characterization of TiAlN Layers

As regards TiAlN, an experimental PVD, Arc Evaporation Reactive Ion Plating plant has been used with two different evaporation methods:

i. simultaneous evaporation of a titanium cathode and aluminium cathode (Figure 1 shows a sketch of the plant),

ii. evaporation of a titanium - aluminium alloy.

Different compositions of high speed steel samples (M2, M42, and ASP30) were coated and characterized. On the basis of both the characterization and "scratch-test" results, the composition Ti=43.7% and Al=56.3% has been chosen as most promising, and the experiments on various tool types are on-going with this composition (Figure 2).

During the TiAlN coating cycle with this first evaporation method, the following problems have been observed:

1. Simultaneous evaporation may give rise to non-homogeneity problems in the deposited-layer composition which are a function of the position of the specimens in the coating chamber.

2. With this method the composition of the TiAlN layer can vary through to the substrate.

The second evaporation method has been used to coat samples made up of 35 CD4 steel with different deposition parameters. To carry out this test, two different titanium-aluminium alloys have been used:

(A) Ti = 75% Al = 25%
(B) Ti = 89.55% Al = 6.38% V = 4.07%.

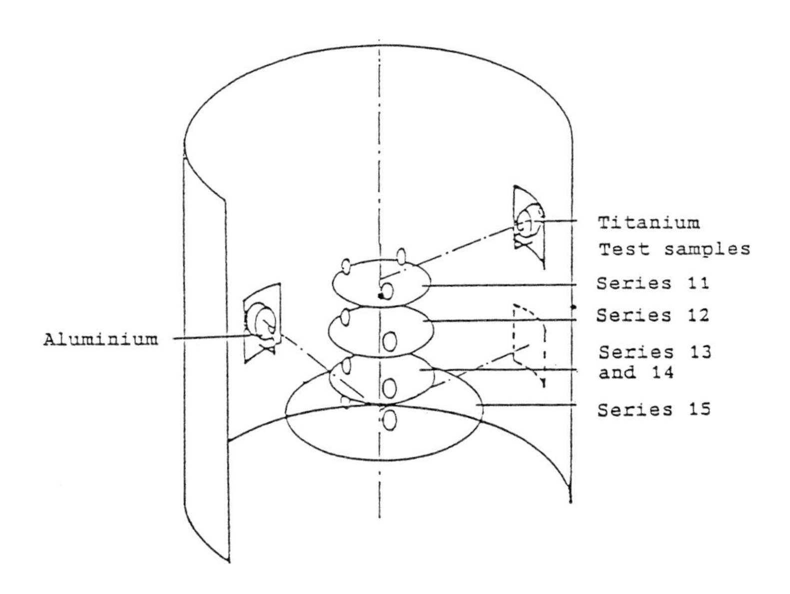

Figure 1 Sketch of PVD plant used to coat samples with TiAlN compound

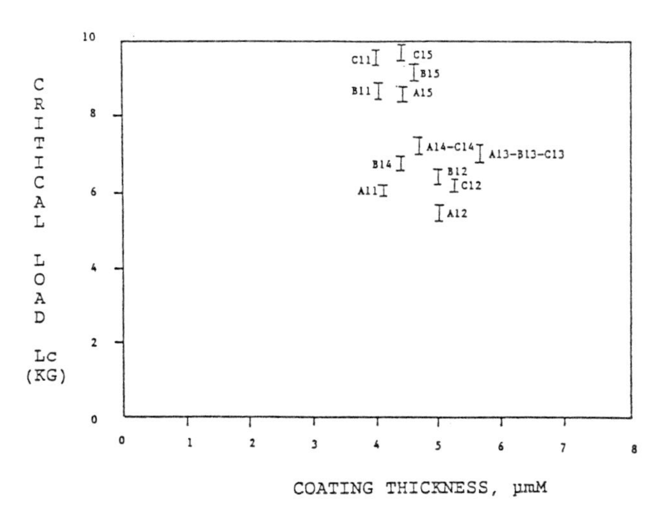

Figure 2 Scratch test on TiAlN coated specimens

During the coating cycles using this second method of evaporation of TiAlN the following problems have arisen:

1. The composition of the TiAlN layers is not the same from the centre of the specimen to the edges.

2. A decrease in the percentage of Al has been observed with an increment of the deposition parameter "polarization bias".

3. The percentage of Al in the deposited layer is independent of the deposition parameter "evaporation intensity".

Formation and Characterization of TiHfN Layers

As to TiHfN the same type of plant with simultaneous evaporation of a titanium cathode and of an hafnium cathode has been used. Hafnium is a rare material that only in the past few years has been used in studies for new coating types. The three different compositions of high speed steel specimens have been tested and then characterized.

On the basis of the results of both characterization and scratch test the composition of Ti = 32.22% and Hf = 67.78% has been chosen as most promising; on going studies are using this coating composition on various tool types - Figure 3.

During the TiHfN coating cycles the following data have been found:

a. The percentage of hafnium in the deposited layer is the same in all samples, independent of their position in the coating vessel.

b. The composition of the TiHfN layer is the same on all points of the substrates.

c. The temperature at the surface of the substrate during the deposition cycle was recorded as 400^{0}C or less, even though in the case of the evaporation of hafnium the intensity of evaporation must be maintained at a very high level (about 80 A).

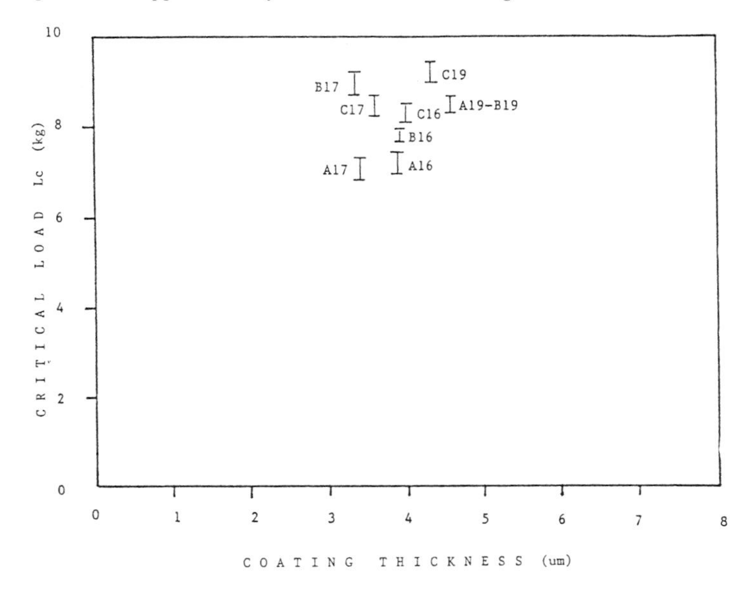

Figure 3 Scratch test on TiHfN coated specimens

3 DEVELOPING AND SET-UP OF DIFFERENT MACHINABILITY LABORATORY TESTS

The objective of this research phase was to define the real benefits of HSS tools coated with the new PVD coatings and with the double treatment of plasma nitriding and PVD process, compared with the conventionally coated HSS tools. The real behaviour of the new coated tools was evaluated by different job-shop machinability tests. During the production tests the machining cycle parameters were optimized for each type of coated tools. After the tests a complete economic evaluation was executed of the benefits of the new coating tools produced.

As machinability tests we have developed three different laboratory tests for the following tools:

A. HSS drills: 8.5mm diameter, esec. N, in HSS M2 for the TiN, TiHfN, TiAlN and plasma nitriding + TiN PVD coatings.

B. HSS end mills: 20 diameter, in HSS M42 for the TiN, TiHfN, and TiAlN coatings.

C. HSS form tools: 40x2.5mm in HSS M2 for the TiN, TiHfN, TiAlN, plasma nitriding + TiN PVD.

The different tools were classed in lots by hardness and cutting edge geometry. Reference lots were obtained from the trade, so as to evaluate any difference in quality due to coating process.

Drilling Tests

These drilling tests were run on an Olivetti Auctor CN drilling machine with different lots of HSS drills with the following characteristics:

- Material: HSS-M2
- Diameter: 8.5mm, esec. N.

In the tests we have considered new tools coated with different coatings and compared these with the standard TiN coated tools. The results of these tests are reported in Figure 4 and show the following indications:

a. The plasma nitriding + TiN PVD process presents the best results (better than the standard TiN coating) only at low speed, and the optimal cutting condition is V_t= 36 m/min.

b. The TiHfN coating did not present good performance under the drilling conditions.

c. The TiAlN coating presents better results than the standard TiN coating only at very high speed, and the optimal cutting condition is V_t = 72 m/min.

Form Tool Tests

These tests were run on an UTITA CNC 22 kW lathe, with form tools of HSS, which best simulate the conditions of use of shaping tools commonly fitted on automatic lathes for large production runs. The tools were tested with the different conditions and compared with the standard TiN coated tools. The test conditions were as follows:

- tool material: HSS-M2
- tool dimensions: Triangle of 40x2.5mm
- work material: UNI C40 steel
- feed: 0.1 mm/rev
- cutting speed: 45-75 m/min

- coolant: 5% emulsion
- tool life criteria: cutting edge failure

The results of these tests are reported in Figure 5 and show the following indications:

a. The plasma nitriding + TiN, the TiHfN and the standard TiN PVD present very similar results in this operation, and the optimal cutting condition is V_t = 55 m/min.

b. The TiAlN coating presents better results than the standard TiN coating when the cutting speed increases and the optimal cutting condition is V_t = 65 m/min.

<u>Milling Tests</u>

These tests were performed on an Olivetti 460 NC milling machine with different lots of HSS end mills with the following characteristics:

- Material: HSS-M42
- Diameter: 20 mm.

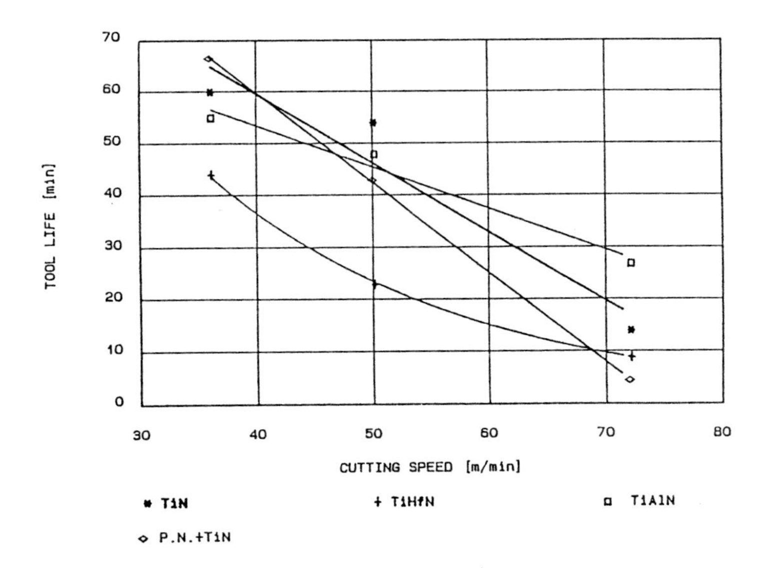

<u>Figure 4</u> Speed versus tool life for the different coatings tested in the drilling operation

In the tests we have considered new tools coated with the different coatings and compared them with the standard TiN coated tools. The results of these tests are illustrated in Figure 6 and show the following characteristics:

a. The TiHfN coating displays better results than the standard TiN coating throughout the range of cutting conditions, and the optimal cutting condition is V_t = 80 m/min.

b. The TiAlN gives similar results to the standard TiN coating, with a small increase of tool life throughout the range of cutting conditions.

Economic Analysis of the Cutting Tests

For the execution of a complete economic analysis of the different solutions tested, we have used the following relation for the determination of the cutting total cost [C] in the milling and form tests.

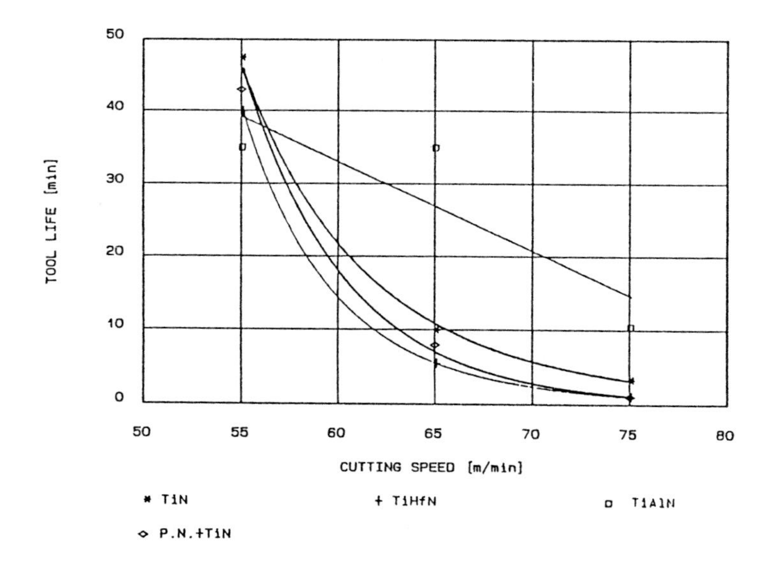

Figure 5 Speed versus tool life for the different coatings tested in the form tool tests

$$C = \frac{(C_0+S_m+C_m) \cdot T_t}{V_{tot}}$$

$$+ \frac{C_u + (C_{aff} \cdot n_{aff}) + (C_0+S_m) \cdot t_{cu} \cdot n_{ut}}{V_{tot}} \quad (Lit./cm^3) \qquad (1)$$

where:

C_0 = cost of labour, Lit/min.
T_t = total cutting time of one tool edge, min.
t_{cu} = tool change time, min.
S_m = depreciation allowance of the machine tool, Lit/min.
C_m = operating cost of machine tool, Lit/min.
C_u = cost of a single cutting edge, Lit.
n_{ut} = number of utilisations of a single edge
n_{aff} = sharpening number of a single edge
V_{tot} = total volume of chip removal during (T_t), cm^3
C_{aff} = cost of sharpening,Lit.
Lit = Italian Lira

The production time is given by:

$$T_{pr} = \frac{T_t + (t_{cu} \cdot n_{tu})}{V_{tot}} \quad min./cm^3 \qquad (2)$$

and the following relationship gives the cost involved in the drilling operation:

$$C = \frac{C_0+S_m}{f} \quad \frac{+ \ (C_0+S_m) \cdot \frac{C_u}{N_u} + C_{aff}}{L} \quad Lit/min. \qquad (3)$$

As examples, we have reported in Figure 7 the curves for the production costs for the coated tools tested in the drilling operations. The economic analysis confirms the results of the production tests and the determination of the optimized machining cycle parameters.

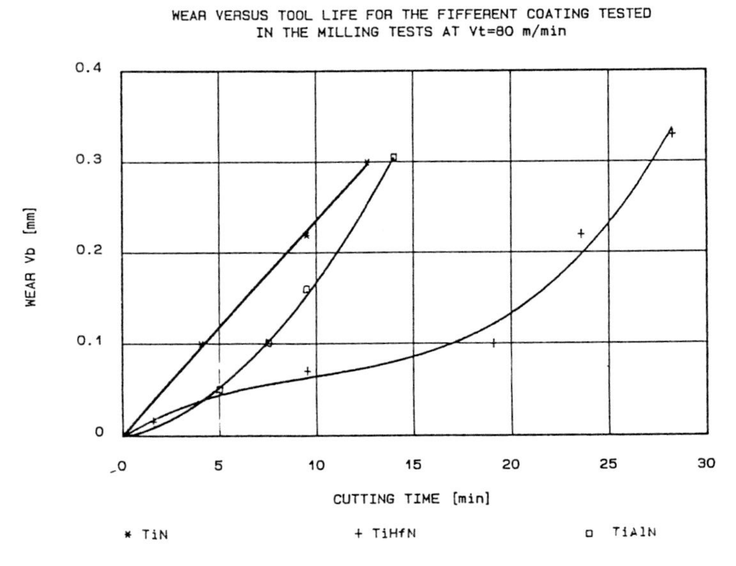

Figure 6 Wear versus tool life for the different coatings tested in the milling tests at V_t = 80 m/min.

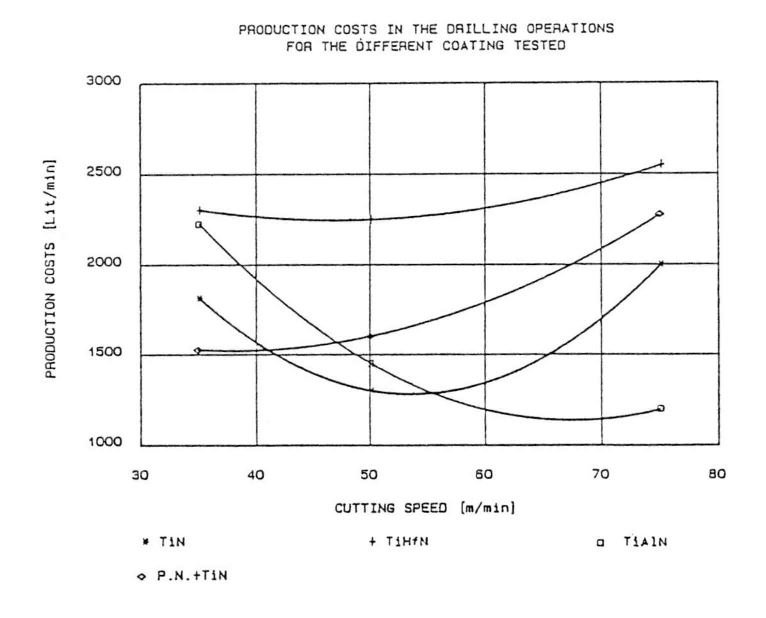

Figure 7 Production costs in the drilling operations for the different coatings tested

4 CONCLUSIONS

1. TiHfN deposits with different stoichiometry have been obtained by using an evaporation source for each metal.

2. The characteristics of the obtained films are generally good and optimal composition of the two metals has been found to be Ti, 32.22% and Hf, 67.78%.

3. TiAlN deposits have been obtained by two different evaporation methods, i.e. using a single source of evaporation of a titanium-aluminium alloy.

4. The characteristics of the films obtained are generally good despite some non-homogeneity defects found during simultaneous evaporation of titanium and aluminium cathodes.

5. In the case of using an evaporating source for each metal an optimum composition of the two metals has been found: Ti, 47.3% and Al, 56.3%.

6. The TiAlN coating is very good for the drill tools and form tools, but only at very high speed.

7. The TiHfN coating is better than the standard TiN coating only for the end mill tools.

8. The plasma nitriding + TiN PVD process presents the best results only on drills and form tools operating at very low speed.

Section 2.5 Quality and Properties

2.5.1
Characterisation and Quality Assurance of Advanced Coatings

K.N. Strafford,[1] C. Subramanian, and T.P. Wilks

SURFACE ENGINEERING RESEARCH GROUP, DEPARTMENT OF METALLURGY, GARTRELL SCHOOL OF MINING, METALLURGY, AND APPLIED GEOLOGY, THE UNIVERSITY OF SOUTH AUSTRALIA, SOUTH AUSTRALIA 5095

[1] CENTRE FOR MANUFACTURING, SOUTH AUSTRALIA 5011

1 INTRODUCTION

Strafford et al[1] have provided a broad definition of surface engineering (SE) in terms of all those techniques and processes which may be used to induce, modify and enhance the *performances* of surfaces with respect to wear, fatigue, corrosion resistance or biocompatibility.

They further identified three major interactive aspects - (i) the design, development and use of surface treatments and coatings, with (ii) optimisation being the key objective, a situation, in theory, achieved, (iii) through full characterisation of such treatments/coatings. It is this last aspect - (iii) - with which this present paper is chiefly concerned - namely how assured quality and desired performance of a surface treatment and especially a coating may be reasonably guaranteed via an adequate knowledge of their appropriate properties and characteristics.

Ultimately, with such a sufficient knowledge and experience of working coating systems it should be possible to design confidently coatings and surface modifications (in certain situations perhaps together on a given component or artifact - duplex surface engineering [2]), via an expert system. Certainly in the area of bulk materials (as distinct from coatings) such methodology has been developed, albeit after many years of research, often at a fundamental level - structure/property relationships - by the materials scientist and engineer.

It is recognised [1] that SE impacts on three major "problem" areas of technology - in the control of wear (tribology), in inhibition of corrosion and in the enhancement of manufacturing efficiency. All of these areas involve the unnecessary expenditure of precious resources - whether financial or material. Also there are major incentives in the associated attraction of added value to products via the adoption of SE practices, as broadly defined.

It has been stated that the term "Surface Engineering" was virtually unknown barely a decade ago, a reference perhaps to the level of sophistication of modern SE practices, particularly with regard to advanced process technologies such as physical vapour deposition (PVD) - ion plating - and plasma - assisted surface hardening becoming available, and the characterisation techniques e.g. XPS, Scanning Auger Spectroscopy etc. However it is useful to recognise that, in fact, SE practices are of a very ancient origin e.g. the development of keen cutting edges on steel tools and weapons (~ 1100 BC). Certainly, commercial electroplating has been in use for nearly two centuries[3], albeit a technique still widely in use in completion with rival technologies and especially the subject of continuous development and increasing sophistication e.g. in its so-called electropulse variant, through which it becomes possible to efficiently plate out amorphous alloys on to artifact surfaces.

The almost bewildering speed with which SE technology is advancing has brought about a considerable gap in knowledge, and especially *understanding* of the precise mechanisms by which observed empirical improvements in artifact behaviour - e.g. the enormous enhancement in manufacturing efficiency associated with the use of tools and dies coated with titanium nitride by PVD techniques - are brought about. Thus in this particular area too little is known of the fundamentals of the low energy plasma physics and the significance and control of process variables - bias voltage, substrate temperature and orientation in the plasma, residual argon pressure, partial pressure of nitrogen etc.[4], and the way in which these determine the nature - physical and chemical properties - of a deposited coating such as TiN. As a result it is not surprising that considerable variability in the performance of coated tools and dies has been experienced. Of equal concern is the lack of understanding of the relationships between coatings' properties and characteristics, and performance, especially in tribological situations. Perhaps a rather better grasp of such features in relation to corrosion performance exists if only as indicated by the many published specifications and quality assurance parameters created in response to the need for reliable performance[5]. However, with the prospects for new novel advanced coatings systems in the context of corrosion control at both ambient temperatures (aqueous and atmospheric corrosion) and elevated temperatures, this situation will also have to be reviewed in order to be able to guarantee optimised performance - "fitness-for-purpose".

Scope of present paper

The present paper is concerned with reviewing the basic properties and characteristics of coatings, and thereby the possible creation of "ideal" coating systems. It will be pointed out however that, in reality, the actual measurement of such properties is in itself often no easy matter, and, furthermore, the significance of such data in relation to performance is often unknown. It is emphasized that **without a much more systematic and fundamental approach to quality assurance of advanced coatings**, there is a danger, at the very least, of a loss of real progress in SE technology; at worst, coatings' users may become disillusioned, reducing the enthusiasm and drive necessary to encourage the research scientist and engineer, and promote the wider active use and adoption of advanced SE technologies.

2 TOWARDS OPTIMISATION IN COATINGS' PERFORMANCE: SOME KEY ASPECTS

2.1 The "Ideal" Coating System - Function and Design.

Figure 1 illustrates, in a generalised way, the working of a coating system - functioning to provide the desired level of control of wear (be it adhesive, abrasive, fatigue, corrosive or fretting[6]), and/or corrosion (be it ambient temperature in an essentially aqueous environment, including atmospheric corrosion, or at elevated temperatures involving oxidation, sulphidation, chloridation etc.[6]).

In simple terms, any coating system - comprising together entities of coating and substrate - are required to relate appropriately to an operating envelope, described as either the working environment or counterface. The former terminology usually is employed in the context of protective coatings and corrosion, the latter with reference to manufacturing processes and wear.

Although the coating and substrate together constitute the coating system it is most important to recognise the marked, distinct, interface which separates these entities - Interface 2 in Figure 1. Interaction between the coating and the substrate across this interface may lead to performance failure - constituted by excessive interdiffusion during fabrication (eg. in the production of a diffusion coating via e.g. chromising or aluminising) and/or during service, especially at the elevated temperatures experienced, for example, in the operation of gas turbine blades[6] or in fast cutting or forming operations. Loss of coating adhesion may occur for a variety of reasons: the coating itself may also suffer loss of cohesion.

More familiar and recognisable are the interactive damage modes or processes which occur across Interface 1, as defined earlier damage modes, in general terms, described as corrosion or wear, the latter taking place either at ambient temperatures (often in aqueous environments - "wet corrosion" -) or at elevated temperatures (often referred to as "dry" corrosion). Wear processes are many and complex but can be satisfactorily classified[7,8] as abrasive, adhesive, fatigue or corrosive, depending on the nature of the environment/counterface - that is, its descriptors or features, and its principal properties or characteristics (as defined in Figure 1), and the coating itself, also similarly defined.

Generally, it is to be noted that the fundamental requirement is *compatibility*, broadly defined. Thus total incompatibility between the substances comprising the coating and the environment/counterface is required in an ideal working system. On the other hand good compatibility between a coating and substrate is necessary to be able to create a sound coating system. It is therefore clear that since these fundamentally different requirements ideally have to be met across both the Interfaces 1 and 2 in Figure 1, no single coating can simultaneously satisfy such requirements, and a degree of compromise has to be achieved in real, practical systems. Nevertheless, in general, adequate systems can be, and are, developed. However, the advent of modern advanced process technologies, perhaps even in combination, offer the prospect of greatly improved and reliable, optimised, performance - "fitness for purpose" referred to earlier - provided adequate proven quality assurance guidelines can be agreed and established. Herein lies the nub of the problem.

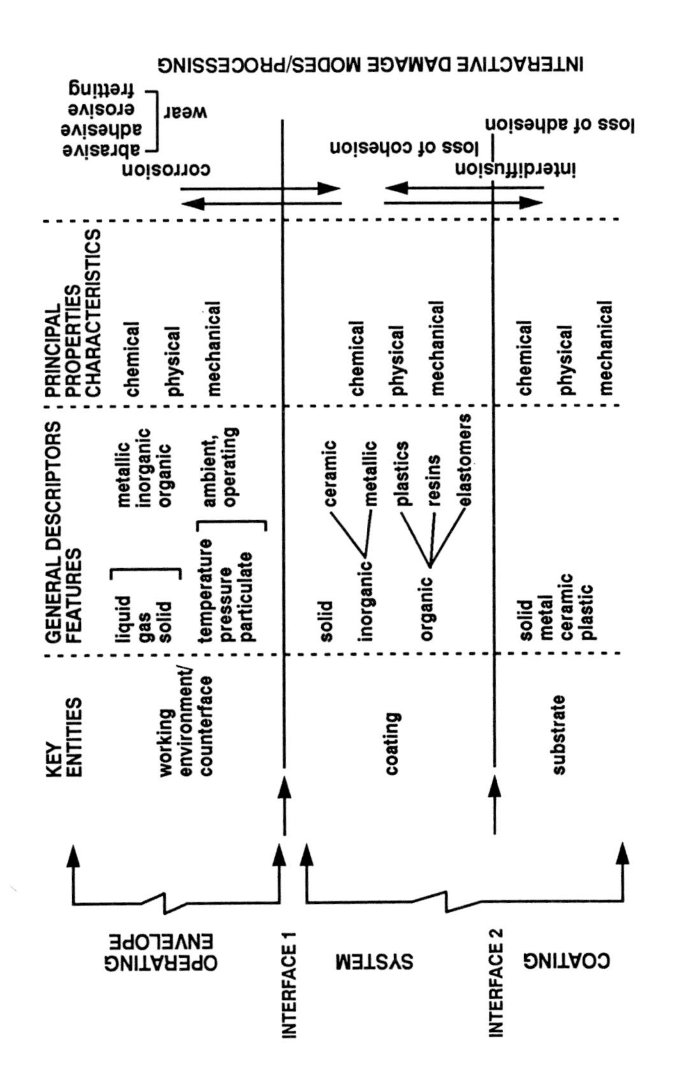

Figure 1 Generalised features of a working coating system.

2.2 Properties and characteristics of coatings - the origins of, and basis for, quality assurance.

The achievement of quality assurance (QA) in advanced coatings - that is, guaranteed performance - fitness for purpose - is governed by four key elements, each, in turn, comprising a large number of individual parameters, largely interactive and interdependent, as summarised in Figure 2[9]. At the fundamental level coatings materials possess a large number of properties and characteristics which obviously ultimately provide the basis of engineering performance - be it tribological (in its many guises), or corrosion. It is the interactive link between these two key elements which is central to coatings design: unfortunately, in general, the precise nature of the linkages between properties and characteristics, and performance, is not understood, although some understanding of the interaction/interdependence has been established, as noted later in Section 3 of this paper.

A key question here then (in the pursuit of the theme of this paper) might be what is the minimum description of a coating in terms of its properties in order to be able to reasonably guarantee performance according to a required life specification. In practice, even with a much better research and knowledge base and the use of expert systems, only limited success is likely to be achieved in precise coatings' design for quite specific purposes - coatings selection. Here the considerable success achieved through systematic research - structure/property relationships - in the field of physical metallurgy/materials science over the last quarter century in the design of bulk materials perhaps however may be noted, and provide grounds for optimism.

Also feeding in as interactive parameters to quality assurance of coatings (Figure 2) are the other two key elements of processes, and process control. A great variety of process technologies are in use, some of which have existed for many years - e.g. electrodeposition as noted earlier - whereas others, such as physical vapour deposition, are of recent origin. Increasingly sophistication in process control is becoming commonplace[10], clearly a crucial element in the production of a repeatable coatings of desired reproducible properties, necessary to achieve the desired performance and function.

Figure 2 Interactive parameters involved in quality assurance of advanced coatings

2.3 The approach to optimised design in a coatings' system.

It is appropriate to review what might be the key property requirements for the development of an "ideal" coating - that is, one which would perform reliably in the chosen working environment in accordance with specification. Strafford et al[9], have fully considered this matter with respect to **tribological performance** elsewhere and only brief reference will be made here to this aspect.

The minimum necessary characteristics have been **generalised** and summarised in Table 1 under three headings - mechanical, physical and chemical properties. The "broad-brush" requirements for the levels values/features of such properties have also been indicated in this Table. While the vagueness and lack of detail of the information given in this Table can undoubtedly be criticized, in reality, little quantitative data exists - some values gleaned from the literature are given in Section 3. Perhaps the most commonly measured parameter is coating hardness and certainly hardness data for many coatings, produced by a variety of process technologies, exist. The significance of such data however in relation to anticipated wear behaviour is questionable and uncertain - precise linkages have not been established. This is also the situation with other coating properties/characteristics cited in Table 1, making the task of designing a coating, *ab initio,* to fulfil **specific service requirements** difficult, if not, impossible.

The crucial importance of the coating/substrate relationship was stressed in Section 2.1 - broadly the compatibility requirements, loosely interpreted, across Interface 2 in Figure 1. A minimum set of "matching" property requirements of coating and substrate, to create an optimised coating system, is given in Table 2. As stressed earlier in Section 2.1, in general, the properties of a working coating and the substrate are, by definition, grossly different and a compromise between the conflicting requirements is needed. Such a compromise, clearly, may be the more-nearly achieved in a multi-layer coating system, offering a gradation in properties from the working Interface 1 through to Interface 2 (Figure 1). Success in this approach has been achieved via the chemical vapour deposition process[11,12], but the flexibility of physical vapour deposition technology, it is suggested offers enormous potential in the design of coatings systems. Within the total design process the ease and precision of modern surface modification processes, such as plasma nitriding and carburising, and ion implantation,

Table 1 KEY PROPERTY REQUIREMENTS FOR THE DEVELOPMENT OF AN IDEAL COATING.

	Properties	Requirements°
mechanical properties	• hardness	high
	• strength (compressive and tensile)	high
	• ductility	moderate
	• toughness	moderate
	• elastic modulus	high
physical properties	• thermal conductivity	high
	• coefficient of thermal expansion	low
	• melting point+	high
	• coefficient of friction*	low
	• density (porosity)	high
	• crystal structure	simple, high symmetry
chemical properties	• chemical compatibility •	good
	• composition	close to stoichiometric

\+ relative to operating or developed service temperature

* relative to counterface-related, *inter alia*, to chemical compatibility, presenting either an inert surface or a surface which degrades at an acceptable rate to yield a corrosion product (commonly oxide) which functions as a lubricant.

• relative to working counterface/environment.

° typical values for proven working coatings materials are given in the text by way of examples.

effecting, for example, "matching" hardness values across Interface 2, should be recognised, Bell[2] has termed this "duplex surface engineering".

In considering the approach to the design **a coating or coatings system for corrosion inhibition** it is necessary to consider other aspects of detail. Key questions to be asked include: why do metals corrode?; how fast do metals corrode (in the chosen working environment)?; what forms/types of corrosion are observed?; how may corrosion control be effected?; how can surface engineering practices be utilised to control corrosion?

Thus, obviously, in choosing, for example, a metal coating to protect a metal substrate there is seen to be a paradox: the coating clearly has to corrode more slowly than the underlying substrate. However, not only has the susceptibility of a given metal to corrosion to be considered, but also the nature of the attack - for example a tendency to pitting rather than more uniform corrosion has to be recognised. In considering how surface engineering may reduce corrosion, the total range of possible methods, including this approach, need to be considered: the use of coatings is merely one, albeit important strategy. Thus generally all of the following methods/approaches may be used/should be considered in corrosion control: selection of materials; design; contact with other materials; mechanical factors; coatings; environment; interfacial potentials. It is to be noted how other aspects include selection of materials and design, factors of major significance in surface engineering, as well as in the particular context of effecting corrosion control.

In determining the use and likely success of protective coatings or surface treatments to inhibit corrosion, some key aspects of corrosion need to be borne in mind, (Table 3). Again, it has been emphasized that the forms of corrosion observed with unprotected bulk metals (Table 4) may also occur with certain metal coatings systems: therefore, a knowledge of the particular susceptibilities of a coating is required at the *design stage,* and careful *selection* of a coating system is required. Again the concept of *"fitness for purpose"* has to be emphasized.

In detail the nature of coatings (Table 5 cf Tables 1 and 2) also has direct implications for the corrosion performance of a coated component. Thus both physical and chemical coating properties are relevant - a heterogeneous porous

Table 2 PROPERTY REQUIREMENTS OF COATING AND SUBSTRATE TO CREATE AN OPTIMISED COATING SYSTEM

Properties	Requirements
• coefficients of thermal expansion	low, matching
• thermal conductivities	matching *
• elastic moduli	high, matching
• chemical compatibility	appropriate to develop high interfacial bond strength and promote alloying+

* thermal barrier coatings should possess low thermal conductivity

\+ a favourable (negative) free energy of coating/substrate interaction should be accompanied by favourable (slow) kinetics of interdiffusion for service in situations where high temperatures are operative or are developed.

Table 3 CORROSION : SOME KEY ASPECTS

- Electrochemical nature of attack
- Origins and types of galvanic cell:
  - dissimilar details
  - metal heterogeneities
  - surface films
  - heterogeneities in electrolyte
- Influence of water composition:
  - pH value
  - dissolved salts
  - dissolved gases
- Influence of flow, temperature and heat transfer

Table 4 FORMS OF CORROSION

- Uniform attack
- Galvanic or two metal corrosion
- Crevice or deposit corrosion
- Pitting corrosion
- Intergrannular corrosion
- Selective leaching - dezincification
  - graphitisation
- Erosion corrosion
- Stress corrosion

Table 5 ON THE NATURE OF COATINGS AND IMPLICATIONS FOR CORROSION PERFORMANCE

- PHYSICAL PROPERTIES
  density (porosity)
  structure
  morphology
- CHEMICAL PROPERTIES
  thermodynamic considerations -
  prediction as linked to likelihood of reaction:
  NO anticipation of speed of reaction: (protective oxide? passivity?)
- (MECHANICAL PROPERTIES)

coating is undesirable, being intrinsically susceptible to corrosion as well as obviously not being able to afford total surface protection: worse, pitting may encourage corrosion of the substrate. Information is required as to the thermodynamic stability of the coating in the environment: however such data will give no indication as to speed of reaction. The coating may initially corrode, only to subsequently passivate and hence afford protection.

Other important factors influencing the corrosion performance of a coating system are set out in Table 6. It is to be noted that porosity in the coating may be open or closed: in the former case, access of the environment to the substrate becoming possible, setting up galvanic effects (which may be harmful or beneficial) or allowing access of gas in a high temperature degradation situation. At high temperatures also interdiffusion effects between the substrate and the coatings are often life-limiting[6]. Table 7 summarises the property requirements for an "ideal" coating/coatings system, something that is rarely totally achieved in reality.

3 OBSERVED LINKAGES BETWEEN COATINGS' PERFORMANCE, AND PROPERTIES AND CHARACTERISTICS: THE EMERGENCE OF QUALITY ASSURANCE STANDARDS

Although it is easy to appreciate the general significance of the portfolio of the properties and characteristics of a coating/coating system in relation to the performance and summarised in Tables 1,2 and more specifically, in relation to anticipated corrosion performance, via Tables 6-8, a detailed and more focussed critical analysis is difficult.

In the first place, as discussed elsewhere by Strafford et al[9], there are experimental difficulties in the actual determination of certain parameters (see Figure 1) of a coating such as Youngs Modulus, Poissons Ratio, coefficient of thermal expansion, thermal conductivity required by the design engineer. As a result such data often do not exist. Again, materials experts traditionally have determined other parameters/characteristics, such as composition, thickness, hardness, adhesion, friction coefficient and wear rate.

Table 6 SOME FACTORS INFLUENCING THE CORROSION PERFORMANCE OF COATING SYSTEMS

- adherence of coating to substrate
- coherency in system - coating/substrate compatibility
- porosity in coating: open or closed
  - gas access?
  - galvanic effects?
- interdiffusion (in HT systems)

Table 7 THE "IDEAL" COATING/COATING SYSTEM

- thin, adherent, coherent (matched physical.mechanical properties)
- chemically inert, or slow rate of degradation (passivation?)
- low porosity
- good galvanic compatibility w.r.t. substrate (if porous)
- good interdiffusional stability w.r.t. substrate

However a major difficulty is to be able to gauge the importance, or otherwise, of such determined data, with reference to actual or anticipated coatings' performance. **In relation to quality assurance this is the nub of the problem.**.

Strafford et al[9], have considered this elsewhere in relation to **tribological performance**, where there are especial difficulties, particularly with reference to the newer advanced coatings such as TiN. In the context of **corrosion behaviour** of coatings the situation is somewhat easier, largely on account of the widespread use of, and experience, in anti-corrosion coatings over many years. However the same cannot be said for the "advanced" coatings - for example hard metal coatings and refractory metals - of potential interest for the inhibition of both wet and dry corrosion, as discussed by Strafford et al[6].

In the literature, quality assurance standards and laboratory (accelerated) methods of testing to simulate performance for anti-corrosion coatings abound. Furthermore these have been in use for many years - see, for example, Carter[5]. Thus, for example, with electrolytic nickel coatings, several standards are available, recommending minimum thicknesses for acceptable service in a variety of atmospheres. It is important to note the developing refinement and speciality of such tests/standards over the years.

In such Standards the thickness agreed to be necessary to provide adequate protection in a given environmental service condition is usually specified as a basic quality control parameter. For threaded components the maximum allowable coating thickness is often also stated. Total coating weight per unit area may also be specified. Because of the inevitable averaging of such data amongst artifacts often of complex geometry (resulting in areas of incomplete or inadequate coverage), it is common to also conduct accelerated corrosion tests to identify such weaknesses e.g. pores in the coating. As always with any accelerated test their use to predict actual environmental performance should generally be avoided, since the degree of correlation which can be achieved is frequently not known or of doubtful validity[5].

It is of interest to note that such well established standards are, in effect, simply comparing corrosion performance of coatings in terms of two parameters -

thickness, and porosity. A thicker anodic coating (e.g. Zn or Fe) will obviously last longer than a thinner one (see Figure 3) : with a thicker cathodic coating (e.g. Ni on Fe) there is much less chance of through-coating porosity, see Figure 4. Thus appropriate galvanic compatibility considerations are assumed, as are the corrosion (or passivation) behaviour of the coating, because of prior knowledge (cf Table 7). Again, although it is well known that, for example, electrolytic chromium or nickel coatings may be in a state of high stress, measurements of the sign (i.e. compressive or tensile) and magnitude of stress are not routinely tested: thus it is a matter of well researched and documented experience that coherency in the coating and adherence to the substrate will be adequate, such parameters being controlled through attention to both formulation and deposition rate, etc. - that is, process control, cf Figure 2.

This situation with well-established anti-corrosion coating systems (it should be recalled, for example, that electrolytic nickel coatings have been in use since 1916) contrasts markedly with that concerning the potential for new coatings/coating systems produced by advanced process technologies e.g. physical vapour deposition (PVD) - ion plating. The difficulties arising here may be illustrated with reference to studies of Strafford and co-workers[6] concerning the potential of titanium nitride and similar "hard metal" coatings, already in widespread use for enhanced tribological performance, as discussed elsewhere[13].

In assessing the potential of TiN as a possible coating for corrosion protection in both ambient aqueous environments at ambient temperatures, and at elevated temperatures (dry corrosion) it is useful and necessary to consider its properties/characteristics in the light of the "check" list provided in Tables 1, 5 and 7.

From a thermodynamic point of view titanium nitride is unstable in oxidizing aqueous environments. The Gibbs free energy data for various reactions at elevated temperatures are given in Table 8[13]. In high temperature oxidizing environments TiN would be expected to undergo oxidation to form titanium oxide or oxy-nitrides. Here it is of special interest to note the positive values calculated for sulphidation reactions - these suggest TiN should be stable in S - bearing environments of sufficiently low oxygen potential. In aqueous situations an important factor could be its galvanic compatibility with respect of a

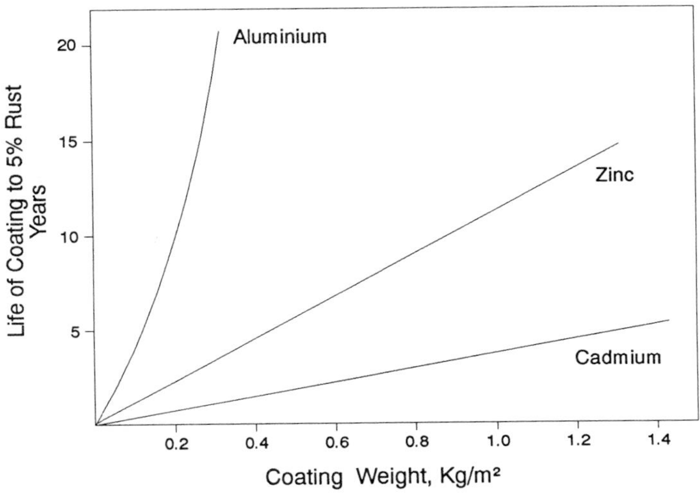

Figure 3 Relationship between coating life and thickness for aluminium, zinc and cadmium on steel exposed in Sheffield. Coatings applied by electrodeposition, cementation, hot dipping and spraying (courtesy: D.R. Gabe in Reference 1 and Ellis Horwood)

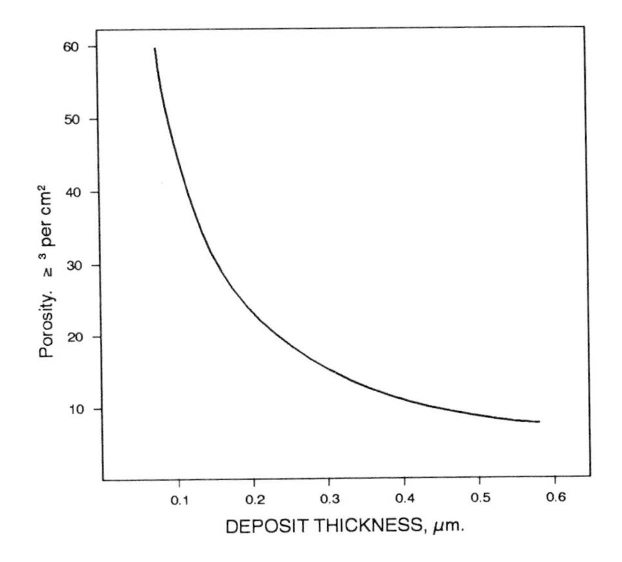

Figure 4 Variation of coating porosity with thickness for electrodeposited chromium (courtesy: D.R. Gabe in Reference 1 and Ellis Horwood)

Table 8 Gibbs free energy data for various reactions

Reaction	$\Delta G°$ (Kcal)		
	500° C	750°C	1000°C
$Ti + S_2 = TiS_2$	-15,838	-5,113	+5,612
$Ti + O_2 = TiO_2$	-183,913	-169,580	-157,624
$2Ti + N_2 = 2TiN$	-126,178	-115,079	-103,979
$TiS_2 + 3O_2 = T1O_2 + 2SO_2$	-312,079	-302,199	-292,319
$2TiN + 2O_2 = 2TiO_2 + N_2$	-110,814	-108,960	-107,259
$2TiN + 2SO_2 = 2TiO_2 + N_2 + S_2$	+35,819	+28,885	+21,909
$4TiN + 2O_2 + 2SO_2 = 4TiO_2 + 2N_2 + S2$	-74,995	-80,075	-85,350
$2TiN + 2S_2 = 2TiS_2 + N_2$	+35,2456	+43,896	+52,546
$TiN + 2H_2S = TiS_2 + \frac{1}{2} N_2 + 2H_2$	+42,532	+40,955	+39,377

substrate e.g. iron or steel. TiN coatings often tend to be porous to a degree and galvanic coupling effects may be critical. The possibility as to whether TiN coatings might become passivated under certain conditions also needs to be considered. Again the integrity of the coating and the nature of the substrate could be critically important under these conditions.

Strafford et al[13], have described the actual corrosion performance of PVD arc evaporated TiN coatings in both aqueous saline solution at ambient temperature, and in oxidising and sulphidising gaseous environments at elevated temperatures. Thus the corrosion behaviour in saline solutions of 43A mild steel substrates coated with different thicknesses of titanium nitride between 2.5 and 10 microns was examined. For comparison purposes other coating types were also prepared namely electrolytic nickel, and electroless nickel phosphorus, aluminium (PVD underlay) and aluminium (PVD overlay) all with 4 microns of titanium nitride. Salt spray corrosion tests were carried out. Polarisation experiments were also conducted to establish whether it was possible to passivate TiN-coated samples. Again comparative tests were conducted with the other coatings systems indicated earlier. High temperature corrosion tests were carried out in air (atmospheric oxidation), in O_2/SO_2 and finally in H_2/H_2S environments. Experiments were performed at 500°, 750° and 1000°C for exposure periods of up to 192 hours.

The weight changes for salt spray treated samples are shown in Fig. 5. All TiN coated 43A mild steel samples exhibited significant weight losses. The samples all showed heavy rust deposits on the surface after exposure to the salt spray. The thicker coatings i.e. 10μ showed less visual corrosion damage to the surface than the thinner ones; for example, with the 2.5μ TiN sample most of the coating had been lost after 48 hr. In general the thicker coatings gave a smaller mass loss and were visibly less pitted. The duplex coating also gave a superior performance compared with the thinner coatings.

The performance of samples produced using alternative coating procedures is shown in Table 9. It is evident that all of these coatings systems perform better than TiN-coated 43A mild steel. In particular the duplex systems exhibit good resistance, particularly that with the Al overlay. Phosphating clearly is not beneficial.

Table 9

Salt Spray Corrosion Tests on Alternative Coating Systems on 43A Mild Steel

(Weight change mg/cm^2, exposure time hours)

Type of Coating				
Time =	24	48	72	96
Electrolytic Nickel Weight change =	-0.17	+0.01	-0.32	-0.38
Time =	24	48	72	96
Electroless Ni + 4μ TiN Weight change =	-0.25	-0.31	-0.24	-0.20
Time =	3	7	24	48
4μ TiN +Al overlay Weight change =	+0.04	-0.05	-0.21	-0.23
Time =	3	7	24	48
Al Underlay + 4μ TiN Weight change =	-0.42	-0.41	-0.47	-0.62
Time =	3	7	24	48
Phosphated 4μ TiN* Weight change =	-0.59	-0.58	-1.36	-3.23
Time =	3	7	24	48
Polished 43A + 4μ TiN Weight change =	-0.17	-0.31	-1.71	-3.95

* Sample was produced by coating 43A mild steel with 4m TiN followed by phosphating.

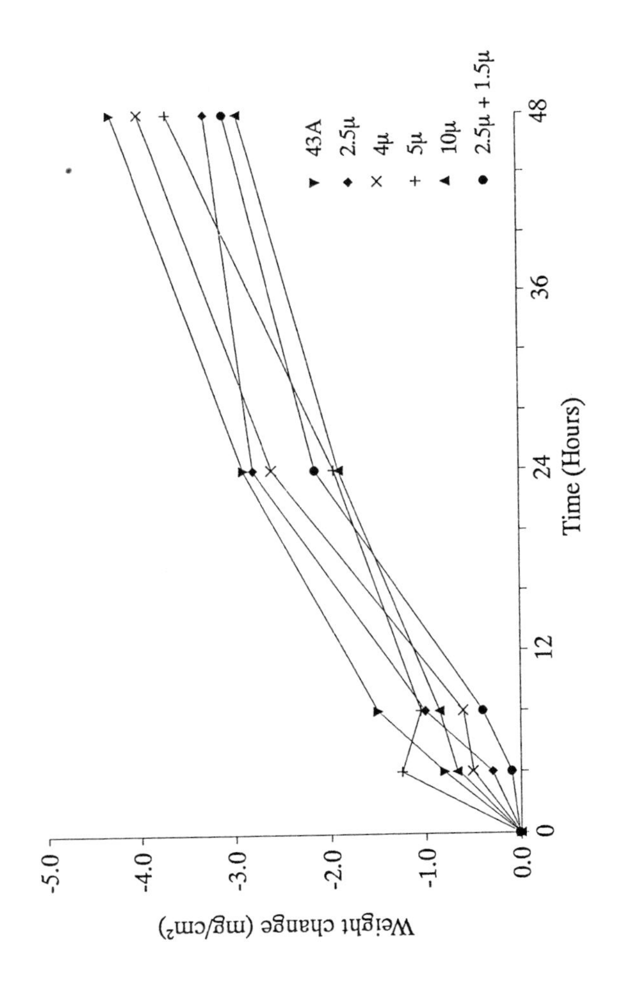

Figure 5 Salt spray tests on TiN coated 43A mild steel

The relatively high mass losses experienced by coated 43A mild steel with only few corrosion pits were interpreted as follows. When corrosion takes place an electrochemical cell is set up around the corrosion site; within the pit anodic reactions take place, causing the dissolution of the substrate, whereas a cathodic reaction takes place on the surrounding area reducing oxygen to hydroxide ions. In this situation the anodic area is relatively small, being the bottom of a pore, and the cathodic area very large, being the TiN coating. The current density in the pore will thus be large causing rapid corrosion. A second effect[14] which is also apparent in pitting corrosion is the enhanced concentration of ionic species within the pit; these will tend to promote electrochemical reactions due to the increased ionic strength of the medium.

Thus a semi-porous film deposited on a susceptible (active) substrate will undergo enhanced pitting corrosion due to this mechanism. Hence to render a substrate corrosion resistant the TiN film deposited must be essentially pore-free, or an intermediate coating which is pore-free and corrosion resistant must be used. **The need to be able to reduce the porosity in TiN coatings through control/manipulation of deposition parameters, including the initial surface finish to the substrate, is clearly crucial in the development of a protective coating system.**

43A mild steel alone does not form a passive layer in the NaCl solution. Although there is a region of the curve (Figure 6), which shows a decrease in current density, the actual current recorded in this region remains too high to be classed as a passive region. Other coated samples gave broadly similar "passive" regions, with the current density being at least an order of magnitude smaller than for the uncoated 43A mild steel. In the case of the two TiN-coated systems the samples were visibly seriously damaged after the experiment i.e. a large number of surface pits were visible. However the electrolytic nickel-coated sample showed damage to the surface.

Although the TiN coated samples show a region of low current density over a wide range of potential, this may not represent the formation of a passive oxide layer. These results of behaviour in salt spray, and studies by other workers[15,16], suggest that TiN itself is an 'inert' material. Studies by Mantyla[16] et al showed a current density of about 10^{-7} A/cm^2 for a dense TiN film in 0.01M

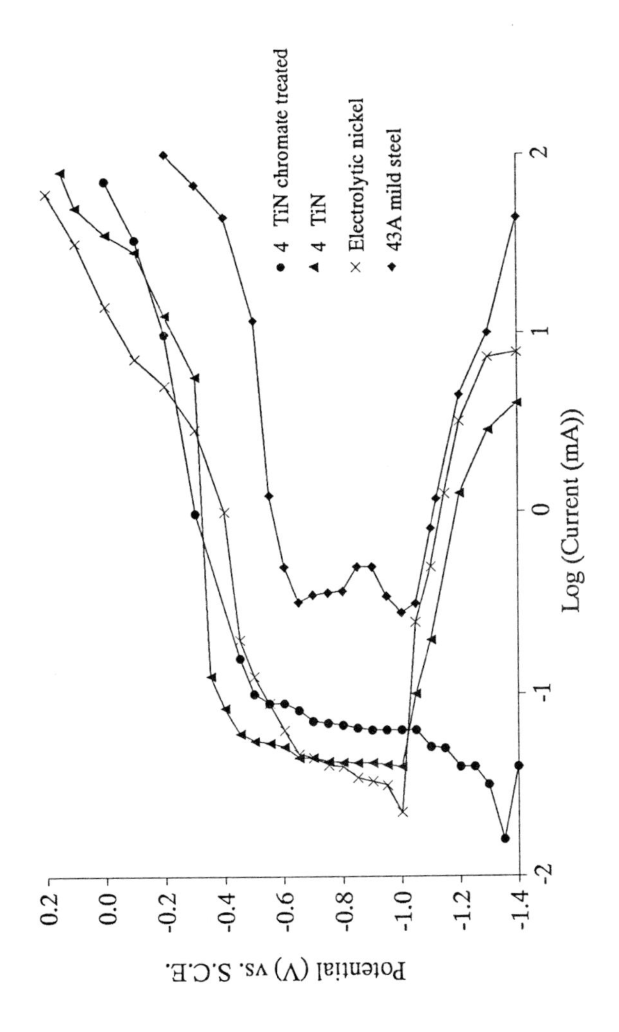

Figure 6 Current-potential curves for various 43A mild steel systems in 3.5% NaCl solution

HCl solution. The authors also stated that the current so recorded may in fact be erroneous due to the insensitivity of their experimental set-up; the true current may be lower than that recorded. For other samples in these studies in which film defects occurred, the linear polarisation curves obtained were similar to the base material, although current densities two orders of magnitude smaller were recorded. In the present case it was felt that the form of linear polarisation curves recorded may be attributed to the partially shielded substrate material, active through an inert semi-porous layer of TiN.

The results of corrosion tests at elevated temperatures on TiN-coated Co20 Cr and Mo substrates in various gaseous environments, described by Strafford et al[13], are also of considerable interest. Thus although TiN is thermodynamically unstable in oxidising environments (Table 8), it was shown to degrade relatively slowly in air at temperatures between 500 and 1000°C. The rate of oxidation of TiN was comparable with the known slow rate of a Co20 Cr control alloy, and much slower than the rate of oxidation of metallic Ti at the same temperature. The amount of corrosion experienced by TiN in an O_2/SO_2 4:1 environment was essentially similar to its performance in oxygen or air. Furthermore the rate of attack was not strongly influenced by temperature: in particular there was no evidence for an accelerated form of attack at reactive temperatures < 750°C, well-known with certain nickel and cobalt based alloys. TiN was indeed essentially inert in a H_2/H_2S environment of low oxygen activity.

4 CONCLUSIONS

Surface engineering practices, particularly those advanced processes such as physical vapour deposition allowing the creation of novel coatings, offer enormous potential in the prospects for enhanced control of tribological and corrosion problems. Thus titanium nitride and other hard metal coatings have demonstrated beneficial influence in greatly enhancing the performance of tools and dies used in a wide range of manufacturing techniques. There is also evidence that such novel coatings may be of use in the inhibition of corrosion.

It has been pointed out that while quality assurance standards are widely available to be able to optimise coating choice and type for best corrosion performance with conventional coatings, generally in the field of advanced coatings, this is an area of neglect. While a portfolio of properties and characteristics of such coatings can be drawn up (to create a basis for the establishment of the assurance of quality), its precise significance in relation to actual coatings' performance is uncertain. Also, a number of these coating parameters are inherently difficult to quantify (measure) and control.

There is evidently a need to conduct systematic basic research into advanced coatings systems - structure/property relationships - and associated process technologies, in order to be able to confidently generate design data to be used in the creation of novel coatings, providing guaranteed optimised tribological and corrosion performance - "fitness-for-purpose".

REFERENCES

1. K.N. Strafford, P.K. Datta and J.S. Gray, "Surface Engineering Practice - Processes, Fundamentals and Applications in Corrosion and Wear", Ellis Horwood, Chichester, U.K., 1990, p21.

2. T. Bell , Metals and Materials, 1991, 7, 478.

3. L.L. Shreir, Ed., "Corrosion" 2 Corrosion Control, p.13.3, Newnes, London, 1976.

4. Reference (1), Chapter 1.2.3., p.89.

5. V.E. Carter, Ed., "Corrosion Testing for Metal Finishing, Butterworths, London, 1982.

6. Reference (1), Chapter 3.1.1., p. 397.

7. J.T. Burwell and C.D. Strang, J. Applied Physics, 1952, 23, 18.

8. C. Subramanian and K.N. Strafford, to be published 1992.

9. K.N. Strafford, C. Subramanian and T.P. Wilks "Properties and Characteristics of Advanced Tribological Surface Coatings and the Assessment of Quality-for-Performance for Enhanced Manufacturing Efficiency". Paper to be presented at Asia-Pacific Conference on Materials Processing, Singapore, 23-25, Feb. 1993.

10. K.N. Strafford. Private communication with T. Bolch, FRI Stuttgart, October 1991.

11. C. Subramanian and K.N. Strafford, to be published 1992.

12. A.S. Gates, Jr., J. Vac. Sci. Technol., 1986, A4, 2707.

13. K.N. Strafford, P.K. Datta and P. Hatto, "The Corrosion Behaviour of Titanium Nitride Coatings". Paper presented at 30th Annual Australasian Corrosion Association Inc. Conf. Auckland, New Zealand, November 1990. In Conf. Proc.

14. T. Suzuki, M. Yamake and Y. Kitamura, Corrosion, 1973, 29, 18.

15. A. Erdemir, W.B. Carter, R.F. Hochman and E.I. Meletis, Mater. Sci. Eng. , 1985, 69, 1.

16. T.A. Mantyla, P.J. Helevista, T.T. Lepisto and P.T. Sictonen, Thin Solid Films, 1985, 126, 275.

2.5.2
Bonding Strength of Films Under Cyclic Loading

R. Gao, C. Bai, K. Xu, and J. He

RESEARCH INSTITUTE FOR STRENGTH OF METALS, XIAN JIAOTONG UNIVERSITY, 710049, XIAN, PEOPLE'S REPUBLIC OF CHINA

1 INTRODUCTION

Hard coatings produced by physical vapour deposition (PVD) and plasma chemical vapour deposition (PCVD) have been widely used in die and tool industries to improve wear and corrosion resistance. An important form of failure of the coated tools and dies is the film peeling from the substrate. Usually, debonding of the film starts from a tiny chip falling off the surface after a number of cycles, then the peeling spot increases in area with an increase in the number of cycles. Therefore, the bonding strength between the film and the substrate should be evaluated with the debonding criterion under cyclic loading in mind. Unfortunately, there is no standard method to test the cyclic bonding strength. Most wear tests are evaluated using weight loss and the wear resistance behaviour of the coated material is observed regardless of whether the bonding strength is good or not. When contact fatigue tests have been used for the bulk materials, the loading stress needs to be high enough to initiate a crack beneath the surface and form pits and scuffs (which are of several tenths of millimetres in depth). The thickness of the PVD and PCVD layer is only a few microns and the normal loading stress for contact fatigue of the bulk sample is generally too high to check the film-substrate bonding strength. Therefore, it is the purpose of this study to find a method for the determination of cyclic bonding strength.

Scratch, indentation, pull-off and other tests have been applied to determine the static and quasi-static bonding strength. The critical loading force L_c in the scratch test is widely used as an index to characterize the bonding strength. However, the film peeling from the substrate, is due to the shear stress in the film-substrate boundary while the L_c value tested with normal press loading is associated with the mechanical behaviour of the film and the substrate, of which elastic and plastic deformation and fracture are involved. Therefore it is hard to detect the shear stress in the boundary layer by the critical loading force[1], even so this might be expected to predict the integrity of the coating layer in real applications, but Valli[2] showed that the cutting life of a tool was not related to the value of L_c. In fact, L_c can only be used as a relative value to compare the bonding strength of the film and the substrate prepared in the same manner. Therefore, it is also a purpose of this paper to study the correlation of L_c and the cyclic bonding strength.

2 EXPERIMENTS AND RESULTS

The substrate material used in this study was a high speed steel (HSS), whose constituents are shown in Table 1. The

Table 1 The constituents of HSS substrate

Elements	C	W	Mo	Cr	V	S	P
wt(%)	0.85	6.32	5.09	4.00	1.98	0.025	0.025

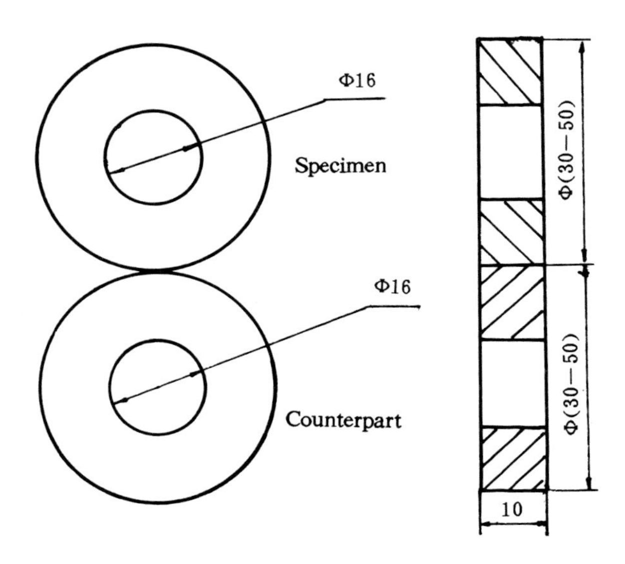

Figure 1 The shape and dimensions (mm) of specimen and counterpart

specimens were coated with TiN film by PVD and PCVD methods.

Method to Test Cyclic Bonding Strength

The specimens of HSS were machined to the shape shown in Figure 1 and heat treated in the normal way to a hardness of HRC 63-65. The counterpart of the roller was made of hardened bearing steel. Two rollers were pressed together in the M-200 wearing machine with a sliding rate of 10%. The normal stress can be adjusted to achieve debonding during cyclic loading. The experimental settings and the choice of parameters are explained

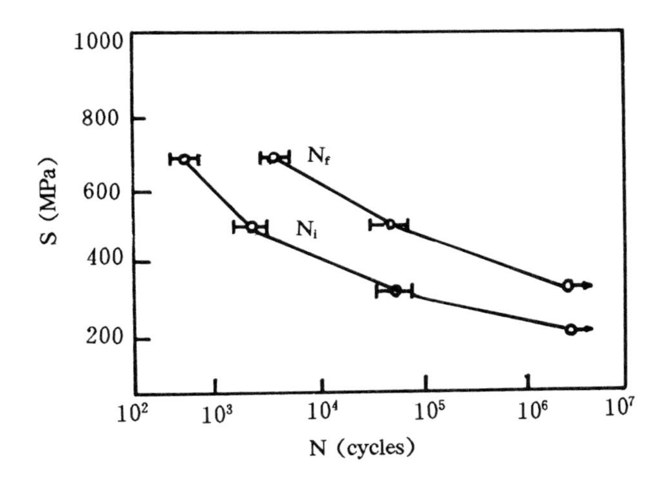

Figure 2 S-N curve for PCVD TiN of 6μm thickness

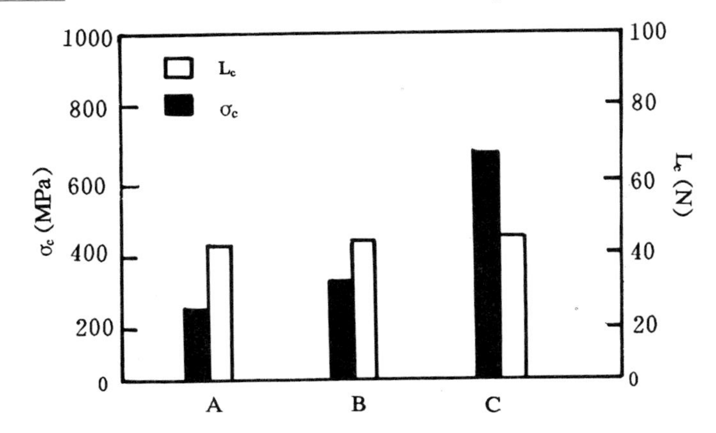

Figure 3 Effect of cleaning process on σ_c and L_c

elsewhere[3]. First, a bare specimen without coating was tested under a normal stress of 750 MPa. It was found the specimen suffered weight loss after $5x10^6$ cycles but no signs of contact fatigue cracks were evident, thus the bulk material did not suffer from pitting- or scuffing-induced contact fatigue. When the coated specimens were run-in the same way with different loading stresses less than 750 MPa, pits appeared on the surface. Auger Electron Spectroscope, Electron Microprobe and Scanning Electron Microscopy were used to analyze the constituents of the materials at the bottoms of the pits. It was found that almost all of the chips peeled from the interfaces of the films and the substrates. This clearly confirms that this experimental method can be applied to determine the cyclic bonding strength.

S-N Curve for the Coated Specimens

If a 5mm diameter pit that appeared on the surface was taken as the initiation of cyclic failure, the cycles to achieve 5mm pitting were named as Ni and Nf cycles indicated the film falling off totally. The experimental data for PCVD TiN of 6μm thickness are shown in Figure 2. If $5x10^6$ cycles are selected as a limit, the contact stress value for 5mm size debonding is defined as the cyclic bonding strength σ_c.

Different Substrate Cleaning Processes

Three pre-coating surface cleaning processes were chosen for cyclic bond strength tests. These were (A) polished surface without cleaning, (B) ultrasonically cleaned in an acetone solution, and (C) ultrasonically cleaned in acetone after sand blasting. PCVD was used to deposit a 6μm thick TiN coating. The experimental data for cyclic bonding strengths are shown in Figure 3. The critical loading forces L_c are also plotted for comparison. The variation in σ_c is significant for different cleaning processes, while the L_c value does not change much.

Effect of Surface Roughness on Bonding Strength

Three states of surface roughness were prepared by polishing, grinding and sand blasting. A 6μm thick coating of TiN was laid down using PCVD. Figure 4 shows the experimental results. Scratch test derived values of

L_c are sensitive to the grinding direction, they being lower in the transverse direction than in the direction parallel to the grinding tracks.

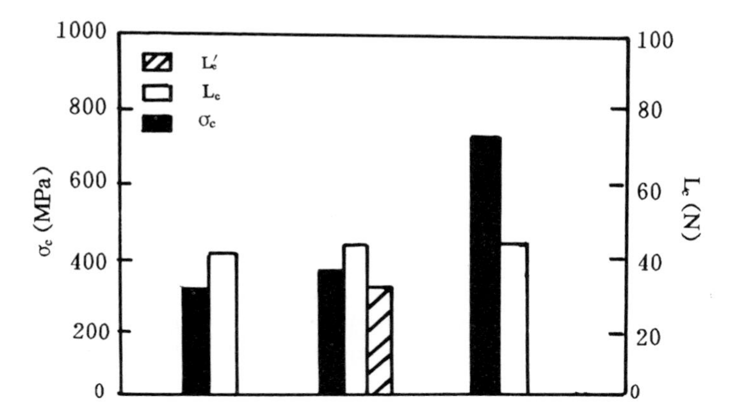

Figure 4 Effect of surface roughness on σ_c and L_c (L_c' - transverse to the grinding direction)

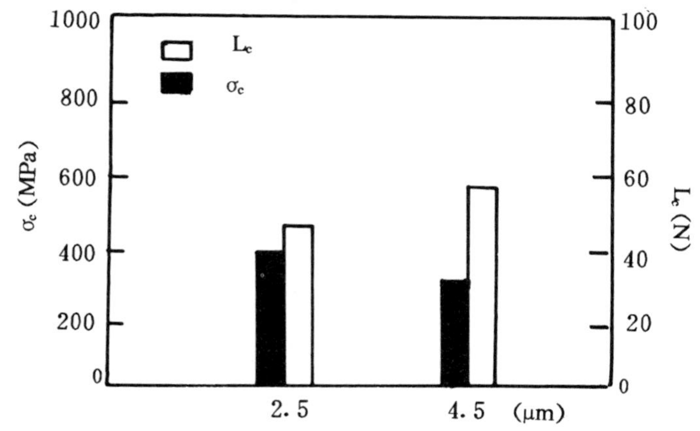

Figure 5 Effect of film thickness on σ_c and L_c

Effect of Film Thickness on Bonding Strength

L_c and σ_c data for PVD TiN coatings of 2.5μm and 4.5μm thickness are expressed in Figure 5 - the value of L_c is higher (conversely σ_c is lower) in the case of the thicker coating.

3 DISCUSSION

Critical loading force L_c in the scratch test only changes slightly for different cleaning processes and roughnesses of the substrates, but the cyclic bonding strength, σ_c changes significantly for different surface states, particularly for the sand blasted condition. The fact that the increment of σ_c after sand blasting is much larger than that after cleaning or grinding agrees well with the experiences in manufacturing. The increase of coating thickness would induce high residual elastic strain energy and lower the bonding between the film and the substrate. The σ_c value decreasing with the increase of the coating thickness reveals the effect of residual elastic strain energy, however the L_c value increases. Generally, the thicker the layer the higher the L_c value from the scratch method, since plastic deformation and fracture are involved, a thick layer associated with a large cross-section for deformation and fracture would exhibit a high resistance to rupture.

The difference of static or quasi-static behaviour and cyclic behaviour for the strength of bulk material has been extensively studied. For low strength material, some correlations can be built between static and cyclic loadings, but for high strength material, the fatigue strength scatters in a wide range for the same static strength, since the material is sensitive to defects or surface roughness. It is the same case for the film and substrate system. Failure under high static stress can be different from that of low cyclic stress, of which damage may accumulate through a number of cycles. The mechanism of fatigue crack initiation and propagation should be different from that of a sudden rupture, likewise the predominant factors are different. The comparison of static and cyclic loading behaviour in the film and substrate system has not quite been addressed, this paper tried to emphasise the difference of these two loading states. However, the cyclical bonding strength test is time consuming and much more difficult than the static test, and it would be beneficial to find some other static criterion for the evaluation of bonding strength.

4 CONCLUSIONS

1. Coated-roller specimens under normal pressure and constant sliding rate can be used to determine the cyclic bonding strength. A 5mm diameter chip peeling from the film-substrate interface is defined as the initiation of cyclic debonding. 5×10^6 cycles are

taken as the cyclic bonding strength σ_c if the debonding area is less than 5mm.

2. For different processing states of substrate surfaces and different thicknesses of PVD or PCVD coating layers, cyclic bonding strength σ_c agrees well with the endurance life expected from manufacturing experience.

3. Critical loading force L_c shows little change for the different processing states of the substrate and increases whilst cyclic bonding strength decreases with an increase in the coating layer thickness.

ACKNOWLEDGEMENT

The authors are grateful to the China National Science Foundation and Lan Zhou Solid Lubrication Laboratory for their kind support.

REFERENCES

1. A.J. Perry, Surface Engineering, 1986, 2,(3), 183.
2. J. Valli, J. Vac. Sci. Technol., 1986, A4,(6), 3007.
3. R. Gao, K. Xu and J. He, J. Vac. Sci. Technol. (In Chinese), to be published.

2.5.3
Characterisation of Thin Metal Films by Nanoindentation

R. Berriche and R.T. Holt

STRUCTURES AND MATERIALS LABORATORY, INSTITUTE FOR AEROSPACE RESEARCH, NATIONAL RESEARCH COUNCIL OF CANADA, OTTAWA, ONTARIO, CANADA K1A 0R6

1 INTRODUCTION

Over the last few years, the so-called depth sensing indentation instruments, also referred to as nanoindenters have become popular for two reasons: 1) the ability to measure both the load and depth of penetration continuously during the test and 2) the high resolution with which these measurements can be made. With such features, indentation instruments have become more versatile than ever. For instance, they have been used to test the adhesion of very thin films (with thickness as low as 20 nm)[1] and to measure the interfacial mechanical properties in fiber reinforced ceramic composites[2]. They can also be used to determine the elastic and plastic properties[3,4] and the creep[5] and fatigue behaviour[6] of various materials in bulk or thin film form. Such testing capabilities are not possible with conventional (dead weight) microhardness testers. Furthermore, in the latter case only one hardness value is obtained per test and the calculation requires the visual imaging of the indentation.

Although depth sensing indentation has gained popularity in the testing of thin films and coatings (with instruments built by different investigators or instruments available commercially), methods for determining the hardness, the elastic modulus and in some cases the adhesion are not yet well established. In this paper, a new method to compute hardness values from measurements of plastic work and depth during indentation tests is presented for both soft (Cu and Al) and hard (Cr) submicron metal films on single crystals of silicon and alumina. The adhesion of these films by scratch testing is also presented.

2 EXPERIMENTAL PROCEDURE

The specimens used during this study are listed in Table 1. They were prepared by evaporating copper, aluminum and chromium thin films simultaneously onto clean and smooth pieces of single crystal silicon and alumina in a vacuum of $5x10^{-5}$ torr. The thicknesses of the films measured by an alpha step profilometer are also shown in Table 1. For background information, bulk Vickers hardness (HV), modulus of elasticity (E) and Poisson's ratio (ν) values for the materials used during this study are given in Table 2.

Indentation and scratch tests were carried out on an instrument which was adapted from the microindenter apparatus at IBM, Almaden Research Center[7]. This instrument has a load and depth resolutions of 16 μN and 0.5 nm, respectively. A more detailed description of this apparatus and its principle of operation can be found in previous publications[7,8].

In the indentation mode, the indenter, either a Vickers (4 sided pyramid) or a Berkovich (3 sided pyramid) is positioned about 0.5μm away from the surface of the specimen. It is then moved with a piezo-electric transducer (PZT) at a speed of 15 nm/s towards the specimen which is attached to a load cell. When the desired load is reached the indenter is withdrawn from the sample at a speed of 20 nm/s. During both the loading and unloading, a personal computer is programmed to continuously acquire data for the load and the depth of penetration as a function of time. The data acquired is later analysed to obtain load vs. total depth of penetration plots (one per test) which can be used to calculate the hardness and the elastic modulus.

During a scratch test, a conical indenter with a cone angle of 90° and a tip radius of 1 μm, is moved simultaneously in two directions. As the indenter is moved towards the specimen by the PZT device, it is also translated along the specimen surface at a speed of 1μm/s for a distance of about 100 μm to make a 100 μm long scratch. During the test, data for the load, the depth of penetration and the scratch distance are acquired and later analysed and plotted as normal load vs. scratched distance, one plot for each test.

Table 1 Specimen identification, film thickness (t) and indenter used for each material (np - not performed, N/A - not applicable)

Specimen	t (μm)	Indentation Tip	Scratch Tip
CUS1 (Cu on Si)	0.35	Vickers	conical
CUA1 (Cu on Al_2O_3)	0.35	Vickers	conical
ALS2 (Al on Si)	0.4	Berkovich	np
CRS2 (Cr on Si)	0.4	Berkovich	np
SI4 (bare Si)	N/A	Berkovich	np
CRS3 (Cr on Si)	0.6	np	conical
CRA3 (Cr on Al_2O_3)	0.6	np	conical

Table 2 Typical properties of materials of interest (Metals Handbook Vol.9)

Property	Al	Cu	Cr	Si	Al_2O_3	Diamond
HV (GPa)	0.6	1.3	15.5	8.0	21	70
E (GPa)	70	130	279	120	524	965
ν	0.345	0.343	0.21	0.42	0.25	0.27

3 RESULTS AND DISCUSSION

Indentation Results

Description of the Load-Depth Plots. Figures 1a and 1b show load/unload plots from indentation tests with a Vickers indenter, carried out to different maximum loads on samples CUS1 (Cu/Si) and CUA1 (Cu/Al_2O_3), respectively. Note that there is a plot for each test conducted. In both cases, the loading curves from different tests overlap, indicating very good reproducibility of the data. The loading part is not a straight line but is curved because as the load increases there is an increase in the area of contact between the tip and the material. Plastic deformation begins from the moment the indenter penetrates the sample. Hence, a transition from elastic to plastic, similar to that obtained during a tensile test, is not observed. The unloading curve, on the other hand, displays an initial linear region due to elastic recovery.

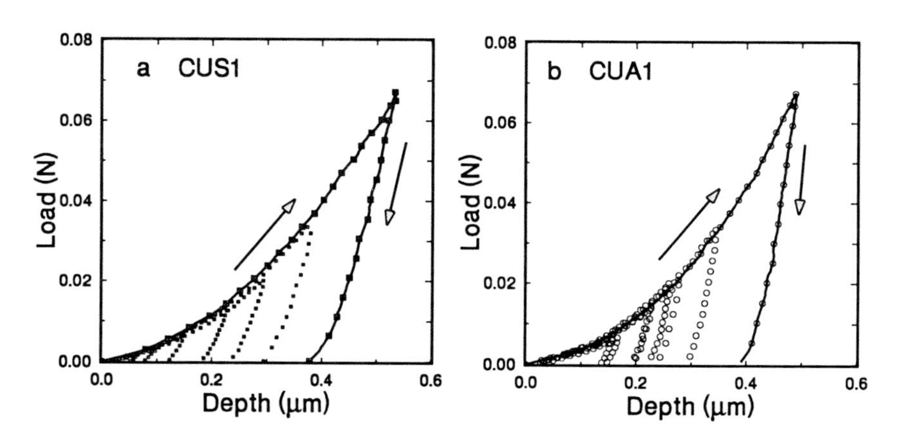

Figure 1 Load vs. depth plots for indentations on samples a) CUS1 and b) CUA1. Arrows indicate the loading and unloading parts

Figures 2a and 2b compare the load/unload plots for samples CUS1 and CUA1 at two different loads. At a load of 0.013 N, there is a perfect match between the two plots because at such a small load the deformation is confined to the copper film and the load-depth curves reflect the elastic and plastic properties of copper without interference from the substrate. However, if the indentation is carried out to a higher maximum load (0.067N, Figure 2b), the two plots overlap at the beginning but deviate as the load increases. Above about 0.025 N, the deformation occurs both in the film and in the substrate and since the sapphire (Al_2O_3) substrate has a higher hardness and elastic modulus than the Si substrate, the plots show that the depth at high loads is smaller for CUA1 than for CUS1. It should be further noted that the divergence between the CUS1 plot and the CUA1 plot occurs roughly at about 0.26 μm which is slightly smaller than the thickness of the film of about 0.35 μm.

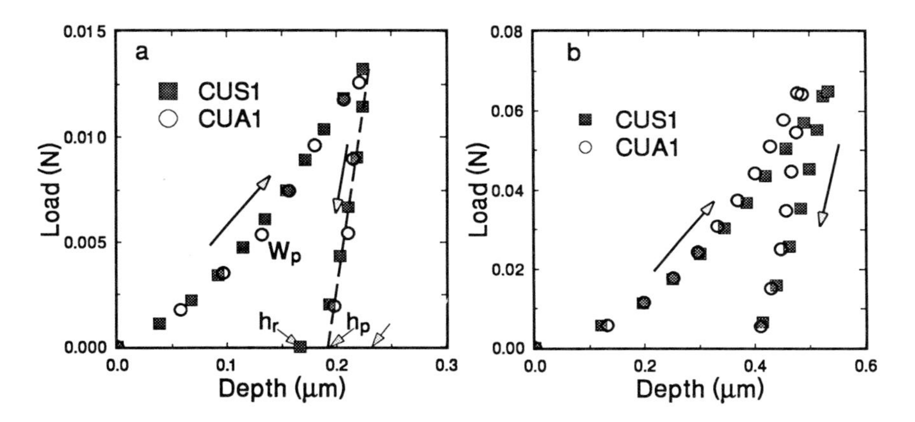

Figure 2 Load/unload plots comparing samples CUS1 and CUA1 at two loads: a) 0.013N and b) 0.067N. Loading and unloading parts are indicated by the arrows

Methods of Data Analysis. Load vs. depth plots, similar to those shown in the above Figures, can be used to obtain the elastic and plastic properties of the thin films as well as to show the effect of the substrate on these properties. The analysis, which is based on the elastic punch theory[9], was first presented by Loubet *et al* [3] and later in more detail by Doerner and Nix[4]. In this analysis the linear portion of the unloading curve is extended to zero load to give the plastic depth h_p - see Figure 2a. h_p can be used to calculate the hardness, and the slope of the line (dP/dh) can be used to determine the elastic modulus (E) of the material being indented. According to Loubet *et al* [3], the expression relating dP/dh and E is:

$$\left(\frac{1-\nu^2}{E}+\frac{1-\nu_0{}^2}{E_0}\right)\frac{dP}{dh}=2\sqrt{\frac{A}{\pi}} \tag{1}$$

where A is the projected area of the indentation, ν is Poisson's ratio and E_0 and ν_0 are the elastic modulus and Poisson's ratio of the indenter material (see Table

2 for approximate values).

In depth sensing indentation the hardness is usually defined as load divided by the projected area to give the Meyer's hardness (HM) (equation 2). Note that HM is about 8% greater than HV, the Vicker's hardness.

$$HM=\frac{P}{A} \tag{2}$$

By substituting for A in equation (2) for both Vickers and Berkovich indenters, assuming ideal geometry for the tip, the expression for HM in GPa with P in N and h_p in µm is given by:

$$HM=\frac{1000P}{24.5h_p^{\,2}} \tag{3}$$

Assuming h_p can be accurately measured, then from equation (3) there is no need to image the indentation impressions to calculate the hardness. However, one difficulty with equation (3) is that it is hard to define the linear portion of the unloading curve for some materials. To standardize the analysis for computer calculations of h_p and the slope of the linear part of the unloading curve, it is suggested that the upper 1/3 of the unloading curve be used as the linear part, regardless of the material (1/3rd rule). It has been found that the h_p values calculated by this method agree with measurements of h_p made manually from the unloading curve.

Since hardness is defined as material's resistance to plastic deformation, the plastic work, W_p, required to form an indentation of volume v is given by W_p=HM.v. The total plastic work, W_p, in µJ is equivalent to the area enclosed by the loading and unloading curves (see Figure 2). Since the geometries, or volumes of the Vickers and Berkovich indenters are equivalent ($v=Ah_p/3$ and $A=24.5h_p^2$), then:

$$HM=\frac{3000W_p}{24.5h_p^{\,3}} \tag{4}$$

thereby, providing a new method to calculate hardness from depth sensing indentation techniques.

Hardness Results. Hardness values for five samples (CUS1, CUA1, SI4, ALS2 and CRS2) were calculated by three different methods:

Method 1 -using equation (3) with h_p determined from the 1/3rd rule.

Method 2 -using equation (4) with the same h_p as for method 1.

Method 3 -using equation (2), with A determined from SEM micrographs.

The results are presented in figures 3a to 3c.

In Figure 3a, the hardness for samples CUA1 and CUS1, computed by methods 1 and 2 were found to increase with depth, which should be expected in cases where the substrate is harder than the coating. However, results using method 3 are independent of the load and are the same for both samples.

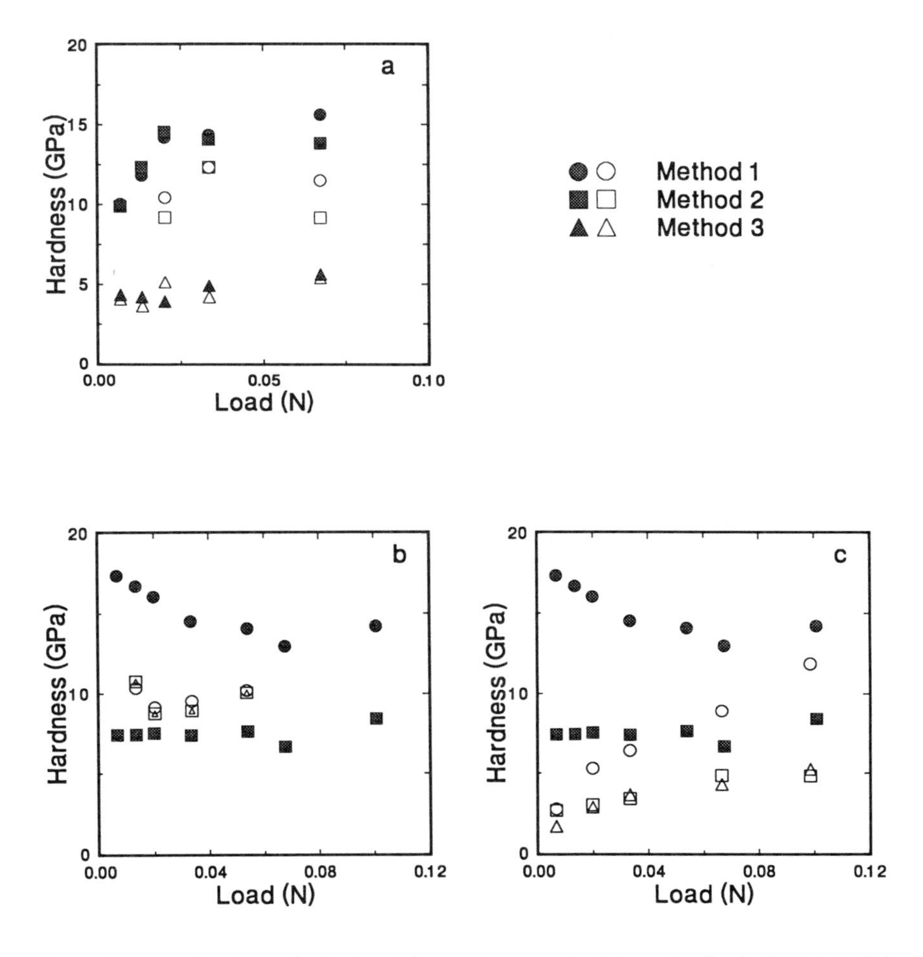

Figure 3 Hardness (calculated by various methods) vs. load. a) CUA1 (solid symbols) and CUS1 (open symbols), b) CRS2 (open symbols) and SI4 (solid symbols) c) ALS2 (open symbols) and SI4

Figures 3b and 3c show the computed values for the hardness of silicon (SI4) and silicon coated with a hard material (sample CRS2) and with a softer material (sample ALS2). The hardness values for silicon calculated by method 2 are very close to the published value (see Table 2) over the entire load range. The hardness values determined by method 1, on the other hand, are much higher, and for some reason decrease with increasing load. For the Cr thin film the hardness calculated by all methods was uniform over the entire load range and, as expected, the value was higher than the value for Si, calculated from method 2. Another discrepancy displayed by method 1 is that it gave higher hardness for Si than for

Cr. For the aluminum coated sample the hardness increased with load, although the results for method 1 were somewhat higher than for methods 2 and 3.

From all these results, it appears that method 2 gave the most consistent and realistic hardness values. Method 3 is not recommended for either soft or hard coatings because a) it gave inconsistent results, b) it is time consuming and c) it is virtually impossible to find indentations produced at small loads during SEM analysis. It should be pointed out that both equations (3) and (4) assume that the tip has an ideal geometry which may not be true for very small depths and consequently may lead to errors. A calibration of tip geometry which involves SEM measurements of indentations in brass, suggested by Stone *et al* [10], was attempted during the present work but was found to be time consuming and ineffective. A better way to monitor tip geometry is to examine the indenter tip in the SEM at appropriate intervals.

Incidently, the remnant depth, h_r (as indicated in Figure 2a) was used in place of h_p to calculate H from equations 2 and 3, and in each case, as Doerner and Nix[4] have shown, the values were much higher than obtained with any of the other methods.

Elastic Modulus Results. According to equation 1, depth vs. load plots can also be used to determine the elastic modulus of the material being indented. The elastic modulus as a function of load is shown for samples CUA1, ALS2 and SI4 in Figures 4a and 4b. The calculations were made using the method of Loubet *et al* [3], modified slightly to accommodate different ν for the materials, i.e. an average ν for the two constituents of the sample. This method, although initially derived for a flat punch, can be applied to other shapes, including the Berkovich and the Vickers tips[11]. The horizontal lines shown in Figure 4 represent published values of elastic moduli. The results show that for the bare silicon sample (SI4) the elastic modulus is independent of the load and is only marginally higher than the value obtained from the literature. As for pure silicon, the calculated elastic modulus for sample CRS2 was found to be constant and again close to the value found in the literature. Since both Cu and Al have modulus values much lower than the substrate (see Table 2), the computed values of the elastic modulus increased with load for both CUA1 and ALS2 samples as shown in Figures 4a and 4b. In the case of CUA1, at the lowest load the calculated elastic modulus is almost the same as the literature value of E for Cu; at the highest load it is almost equal to the elastic modulus of sapphire and at intermediate loads it is between these two values.

Based on these results, we are confident that the values of elastic modulus calculated from depth sensing indentation measurements can be used to characterize thin metallic films. Discrepancies between the calculated values and the actual values may be due to several factors, but mainly because the compliance of the machine has not been corrected for. Additionally, errors due to the geometry of the indenter tip and the anisotropy of the material may be expected. A possible improvement to this analysis is to calculate the elastic work from the load/unload plots and use this value to determine E. The validity of this last method is currently being investigated.

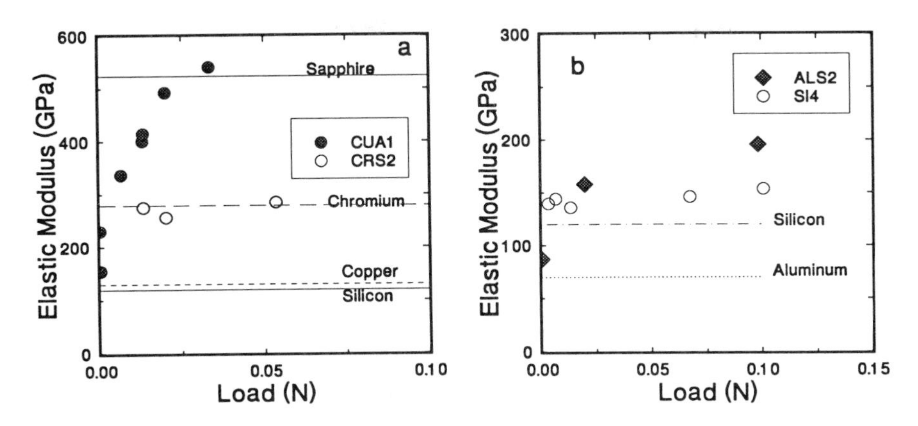

Figure 4 Calculated elastic moduli as a function of load a) for samples CUA1 and CRS2 and b) for samples ALS2 and SI4. The horizontal lines are published elastic moduli

Scratch Test Results

The purpose of the scratch test is to determine the critical load, L_c, needed to cause spallation of the coating. This load can be used directly as a measure of adhesion, i.e. the higher the critical load the better the adhesion. Alternatively, the critical load can be used to calculate a shear stress at failure which is then taken as a measure of interfacial strength (or adhesive strength)[8,12]. With depth sensing indentation, L_c is determined from load vs. scratched distance plots as the first point at which a sudden drop in the load occurs. The scratch may be examined in the SEM to verify spallation of the film.

The load vs. scratched distance plots for samples CUS1 and CUA1 (Figure 5a) exhibit three distinct regions; an initial linear region during which the deformation caused by the moving indenter is still confined to the film, then a transitional region where the deformation occurs in the film as well as in the substrate and a third region where the contribution from the film becomes negligible. In each case the film thickness was 0.35 μm which is the depth reached after a scratched distance of about 25 μm. None of the scratch test plots for samples CUA1 and CUS1 exhibited any load drops which means that the copper film adhered very well to both substrates. This was confirmed by SEM analysis of the scratches (Figure 5c) which showed that the conical indenter tip sequentially cut through the film and the substrate, with no apparent spallation of the film.

Unlike the previous two samples, the load vs. scratched distance plots for the Cr coated samples (Figure 5b) exhibit sudden load drops. For both CRS3 and CRA3 about six scratch tests were carried out and the first load drop, i.e. L_c, occurred exactly at the same load in each case. SEM examination (Figure 5d) of sample CRS3 indicated that spallation of the coating corresponded to the first load drop. The critical load for sample CRS3 is lower than that for sample CRA3. Since both samples have the same coating thickness and since the data is very reproducible, this difference in critical load indicates that chromium has better adhesion to alumina than to silicon.

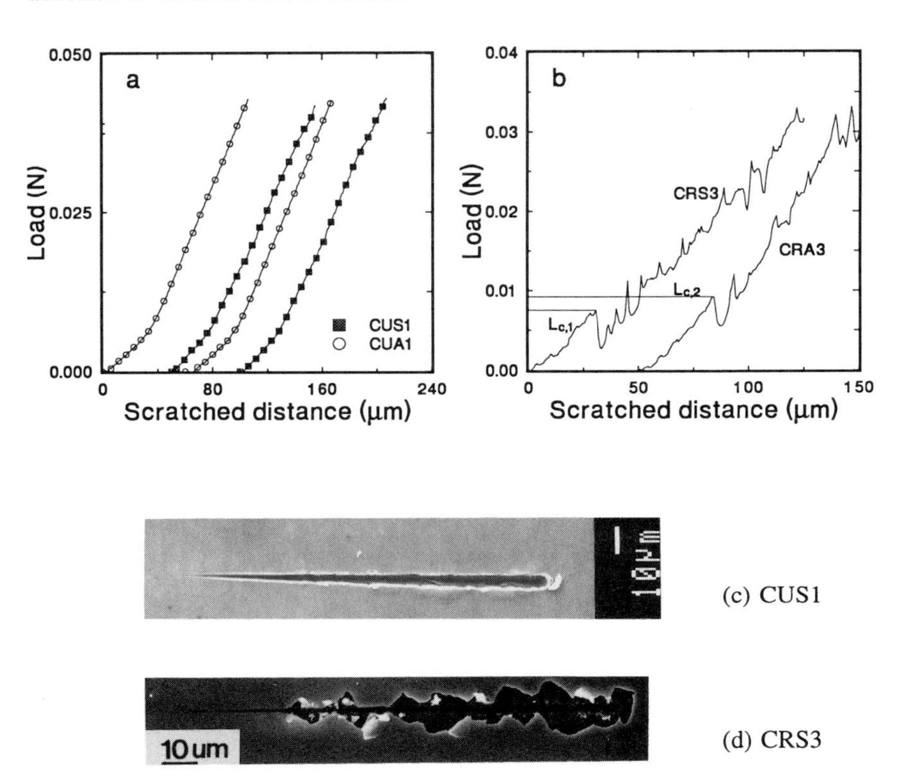

Figure 5 (a) and (b) Scratch test plots and (c) and (d) SEM micrographs of scratches for Cu and Cr films on Si and Al_2O_3 substrates. Note: scratch plots have been displaced along the x-axis for clarity

4 SUMMARY AND CONCLUSIONS

1. Depth sensing indentation tests have been used to determine the hardness and elastic modulus of thin metallic coatings deposited on silicon and alumina single crystals.
2. Of the three methods used to compute hardness values, a newly developed method based on plastic work (method 2) gave the best results over the range of depths studied, conforming to published values for the coating and

the substrate.

3. Values for the elastic modulus of the coating/substrate composite computed from the load-depth plots were also in good agreement with published values.
4. Coating adhesion was derived from reproducible scratch tests by measuring the critical load required to cause spallation (as verified by SEM observation).
5. It was found that hard Cr coatings had better adhesion on alumina than on silicon.
6. Copper coatings on both substrates survived the scratch test to depth well in excess of the coating thickness.

ACKNOWLEDGEMENT

R. Berriche would like to thank Prof. D.L. Kohlstedt for the use of laboratory facilities at the University of Minnesota. The data analysis was carried out at the Structures and Materials Laboratory of the National Research Council of Canada in support of projects LH105 and JHJ00 being carried out in collaboration with Cametoid Ltd. and Liburdi Engineering Ltd., respectively.

REFERENCES

1. J.B. Pethica, R. Hutchings and W.C. Oliver, Phil Mag. A, 1983, 48, 593.
2. D.B. Marshall and W.C. Oliver, J. Appl. Phys., 1987, 70(8), 542.
3. J.L. Loubet, J.M. Georges, O. Marchesini and G. Meille, J. of Tribology, 1984, 106, 43.
4. M.F. Doerner and W.D. Nix, J. Mater. Res., 1986, 1, 601.
5. V. Raman and R. Berriche, J. Mater. Res., 1992, 7, 627.
6. T-.W. Wu, A. L. Shull and R. Berriche, Proceedings of the First European Conference on Diamond and Diamond-like Carbon Films, A. Mathews and P.K. Bachmann, eds. Switzerland 1990.
7. T.S. Wu, C. Hwang, J. Lo and P. Alexopoulos, Thin Solid Films, 1988, 166, 299.
8. R. Berriche and D.L. Kohlstedt, pp.47-54 in Materials Developments in Microelectronic Packaging: Performance and Reliability. Edited by P.J. Singh. ASM International 1991.
9. I.N. Sneddon, Int. J. Eng. Sci., 1965, 3, 47.
10. D. Stone, W.R. LaFontaine, P. Alexopoulos, T.-W. Wu and C.-Y. Li, J. Mater. Res., 1988, 3, 217.
11. G.M. Pharr, W.C. Oliver and F.R. Brotzen, J. Mater. Res., 1992, 7, 613.
12. P. Benjamin and C. Weaver, Proc. R. Soc. London, 254A(1960) 177.

2.5.4 Microstructure/Property Relationships for Nitride Coatings

S.J. Bull and A.M. Jones

AEA INDUSTRIAL TECHNOLOGY, HARWELL LABORATORY, DIDCOT, OXFORDSHIRE OXI I ORA, UK

1 INTRODUCTION

The increasing use of thin surface coatings in many engineering applications brings with it a need for a fundamental understanding of their properties if the optimum coating for a particular purpose is to be selected. Such coatings can be produced by a range of deposition technologies which can be broadly split into two generic types, Chemical Vapour Deposition (CVD)[1] and Physical Vapour Deposition (PVD)[2]. In order to produce coatings tailored for a specific application a knowledge of the interaction between the process or system parameters for these technologies and the microstructure and properties of the coating needs to be determined[3].

The structure and properties of titanium nitride coatings have been extensively studied for a number of deposition technologies[4-8]. The size and packing density of the columnar units which comprise the coating and the strength of the boundaries between them are the important microstructural features which dictate performance in many applications and these depend on the choice of coating and substrate, the temperature of deposition and the energy and flux of ion bombardment which takes place during coating[3].

Since there are many commercial deposition technologies for producing a single coating material it is necessary to define the selection of a coating in terms of property or performance criteria which each process technology has to meet. Ultimately these are related to coating microstructure and thus it is important to determine the microstructure/property relationships for each coating material and how these are related to the choice of deposition technology and process parameters[3].In this paper we consider coatings produced by only one PVD process, Sputter Ion Plating (SIP)[9,10] and compare the previously published microstructure/property relationships for SIP titanium nitride coatings to those for other nitride materials.

2 EXPERIMENTAL DETAILS

The nitride coatings were produced by Sputter Ion Plating (SIP)[9-12], which is a soft vacuum dc sputtering process developed at the Harwell Laboratory of AEA Technology. Stainless steel disc substrates were polished to a 3 μm diamond polish using standard metallographic techniques, ultrasonically cleaned and vapour degreased prior to coating. Single-crystal silicon wafers ((111) orientation) were also used as substrates in some cases. Coatings were deposited over a range of substrate bias voltages (0 to -120 V) for a constant deposition time at a temperature of approximately 500°C. Coating thickness was determined by a ball-cratering technique. Titanium, zirconium, chromium and tantalum nitride coatings were produced by reactive sputtering using a mixture of argon and nitrogen as the sputtering gas; the gas composition was adjusted to give films containing the mono-nitride only.

Vickers hardness measurements were made on all coating surfaces and the stainless steel substrate. The hardness of the coating material was then determined from the composite coating-substrate hardness using a volume law-of-mixtures hardness model[13-15]. The hardness of the coating is expressed as the value at a given indent size (10 μm diagonal) thereby minimising the effect of the variation of hardness with indentation size which is observed in ceramic materials.

Scratch adhesion testing was performed using a commercially available scratch tester (from CSEM, Switzerland) fitted with a Rockwell C diamond stylus (cone apex angle, 120°; tip radius, 200 μm). This is a dead-loaded instrument and hence a number of scratch tracks were made at increments of 200 g normal load until an adhesion-related failure is detected at regular intervals along the length of a single track. The critical load L_c for coating detachment was determined on the basis of spallation/buckling failures ahead of the moving stylus[16] as identified by post facto scratch track analysis by reflected light microscopy. The origins of this and other failure mechanisms are discussed elsewhere[15,17].

Sliding wear tests were performed on a sphere-on-disc tester in unlubricated conditions and lubricated using a rotary pump oil (Edwards 15). An uncoated 52100 bearing steel sphere (3mm diameter) was used as the counterface in all tests. The load at the sphere was ~2 N (~0.9 GPa contact pressure); sliding speed was ~0.1 m/s; total sliding distances were ~100 m (unlubricated) and ~400 m (lubricated). Coating wear was determined from a surface profilometer trace over the wear track cross-section as discussed elsewhere[18,19].

Residual stresses were measured by two techniques. On stainless steel substrates, stresses were determined by x-ray diffraction using the $\sin^2\psi$ technique[20] assuming a biaxial stress state. The method has been described previously[21-23]. On single-crystal silicon wafer substrates, the curvature of the wafers was measured before and after coating. From the change in curvature, thickness of film and substrate, and the elastic properties of the wafer the stress in the coating can be measured, using the Stoney equation[24-25].

3 RESULTS

For all the nitrides investigated the appearance of the coatings was similar. At low bias voltages an open porous film was produced with a very rough surface. In general the thickness (and roughness) of the coating decreases as the bias voltage increases (see Table 1). The TiN and ZrN films were very much thinner than the CrN and TaN films and showed much less variation in thickness as a function of substrate bias.

The 10 μm coating hardness increases with substrate bias voltage for all four of the nitride coating materials examined. The ZrN coatings were hardest and their hardness increases over the complete range of biasing, whereas for TiN the hardness levels off at the higher voltages due to yielding in the coating. Both TaN and CrN show a peak in hardness at about -90 V;for these latter materials there are two possible explanations for this behaviour. Firstly the very high residual stresses in the coatings have caused the substrate to yield and thus relieve some of the stresses in the coating. This results in a permanent bending of the substrate and a reduction in measured hardness. The second explanation is that at high levels of ion bombardment the coating nucleates and grows with a different texture which grows faster into the vapour flux than it does laterally leading to lower coating density and hence reduced hardness. The first explanation is most likely here though the change of growth habit cannot be ruled out for the CrN coating.

The scratch test results for the four nitrides are much less consistent (Figure 2). The adhesion of CrN seems to be very poor and independent of bias voltage. Coating adhesion is improved on this for the TaN and ZrN thin films without forming a definite pattern as the substrate bias voltage changes. The TiN coating displays a trend of

Table 1 Thickness of SIP nitride coatings as a function of substrate bias

	Coating Thickness (μm)	
Coating	Unbiased	-120V bias
TiN	2.9	2.4
ZrN	3.15	1.75
CrN	10.0	3.6
TaN	11.7	5.1

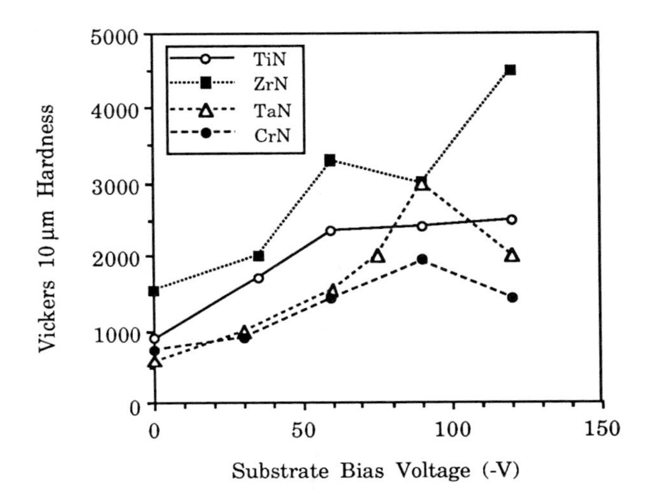

Figure 1 Variation of coating material hardness with substrate bias voltage for SIP nitride thin films

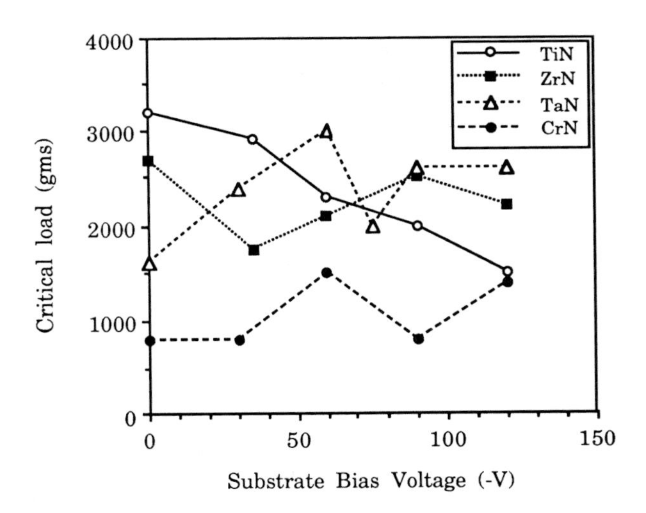

Figure 2 Coating adhesion critical load vs. substrate bias voltage for SIP nitride thin films

a decrease in coating adhesion as the negative substrate bias voltage increases. Direct comparisons between the critical loads as a measure of adhesion cannot easily be made for these materials without taking into consideration the thickness of the coating. In general the critical load increases as the coating thickness increases and thus only the TiN and ZrN results can be compared directly. The low values of critical load for the TaN and CrN films despite their greater thickness shows that the adhesion of these coatings is relatively poor. This is not surprising since no special attention was paid to developing cleaning pretreatments for these materials whereas the treatments have been optimised for TiN and the behaviour of ZrN is expected to be very similar due to the close relationship of titanium and zirconium in the periodic table.

In unlubricated sliding , generally, higher wear rates are seen at the extremes of bias voltage (0 V, -120 V) with minimum wear rates for each nitride at some intermediate substrate bias level (Figure 3(a)). The differences between maximum and minimum wear rates were over three orders of magnitude in some cases. TiN and ZrN coatings showed definite minima in wear rates, at -60 V and -35 V respectively. The thicker CrN and TaN coatings show much lower variations in wear as a function of substrate bias.

Sliding wear rates were generally two orders of magnitude lower in the case of lubricated test conditions (Figure 3(b)). Wear rates were similar over all substrate bias voltages, with only ZrN showing a definite trend – a decrease in wear rate as the negative substrate bias voltage increases.

Residual stress measurements were made only on two of the coating materials, TiN and ZrN (Figure 4). In both materials, the trend is the same, with low compressive stresses at low bias voltages increasing to high compressive stresses as the bias increases.

4 DISCUSSION

Coating Microstructure

In vapour deposition processes coatings are formed from a flux of atoms that approaches the substrate from a limited range of directions and as a result of this the microstructure of the coating is columnar in nature. Thin film microstructures were first classified by Movchan and Demchishin[26] in terms of three distinct structural zones as a function of the homologous deposition temperature and this was modified to take into consideration the presence of a sputtering gas by Thornton[27]. The coatings deposited without a substrate bias have the low temperature zone-1 microstructure which consists of tapered columns with domed tops and is determined by conditions of low adatom mobility. As the surface temperature is increased, surface diffusion becomes more important and the microstructure of the coating consists of parallel-sided

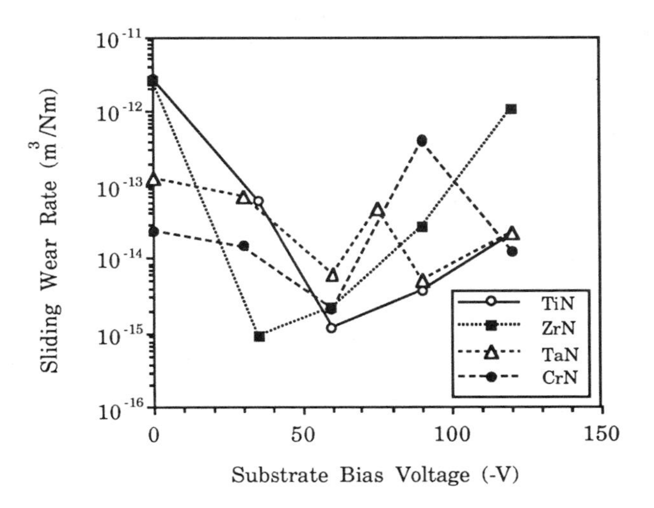

Figure 3(a) Sliding wear rates vs. substrate bias voltage for SIP nitride thin films measured in unlubricated testing conditions

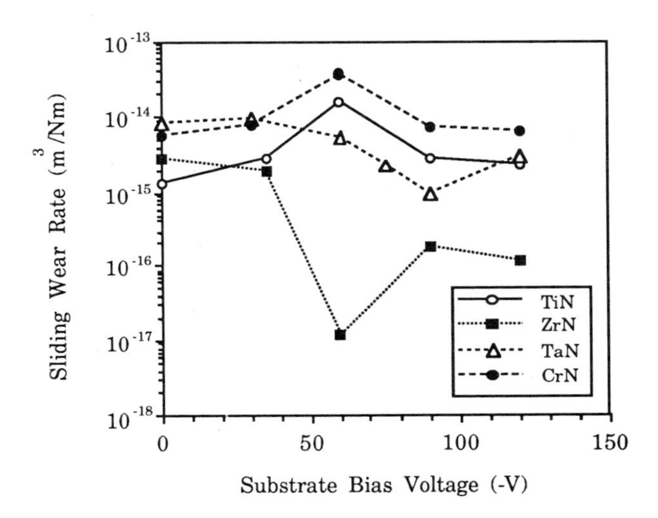

Figure 3(b) Sliding wear rates vs. substrate bias voltage for SIP nitride thin films measured in lubricated testing conditions

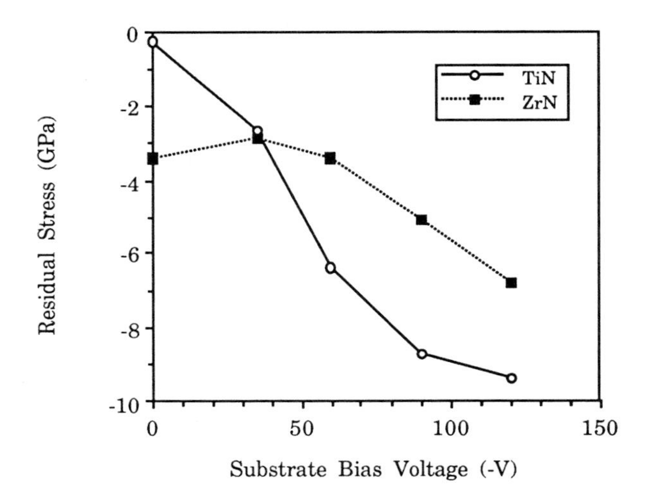

Figure 4 Residual stress vs. substrate bias voltage for SIP nitride thin films

columnar regions which have a smooth surface topography (zone-2). A transition structure, zone-T which consists of poorly defined fibrous grains is often observed for sputtered coatings. At the highest temperatures, bulk diffusion also becomes important and the zone-3 microstructure consists of equiaxed grains.

The application of a negative substrate bias voltage in SIP leads to bombardment of the films with particles which can have sufficient energy to cause the filling of the voided boundaries by coating atoms leading to a denser structure of the the zone-T type[3]. This filling can take place by the knock-on of already deposited atoms, but is mainly due to bombardment-induced mobility of these atoms. All the nitride coatings here showed a transition from zone-1 to zone-T as a function of substrate bias. Though the energy of the bombarding ions is controlled by the substrate bias, the flux which the substrate receives depends on the ion current which is a function of the pressure in the system as well as the plasma intensity at the substrate. Adatom mobility and hence microstructure is a function of the momentum transfer from the ions which depends on both ion energy and flux.

Microstructure/Property Relationships

For zone-1 microstructures the open columnar boundaries ensure that any residual stresses in the deposit are kept to a minimum[28,29]. A direct consequence of the denser microstructures formed at higher bias voltages is that substantial stresses can be generated within the coating. Indeed the apparent increase in density of PVD coatings when moving from zone-1 to zone-T microstructures can also be correlated with the levels of residual stress present in the coatings[28].By virtue of the open microstructure and the subsequent reduced load-bearing capacity of the surface, the hardness of the coating is dramatically reduced over that of the more dense zone-T microstructures formed at higher bias voltages. The hardness of these high bias coatings is comparable to that of bulk materials. However, it should be realised that the hardness of the individual columns will be almost identical for the zone-1 and zone-T microstructures.

For the titanium nitride coatings investigated here the effective adhesion of the coating is reduced as the substrate bias increases due to the build-up of residual stress in the coating. Similar results are expected for the zirconium nitride coatings investigated here but the experimental results presented in Figure 2 indicate that there is no systematic reduction in adhesion. The zirconium nitride films are tougher than the titanium nitride films (as seen by reduced cracking in scratch tracks performed at identical loads) and tend to show different failure modes in the scratch test. It is clear that the failure modes for ZrN and the other nitride coatings are less susceptible to residual stress from the results obtained in this study. For the CrN and TaN layers this is probably due to the much greater thickness of the coatings tested.

The increase in hardness on going from zone-1 to zone-T microstructures does initially lead to an increase in wear resistance[30] which may explain some of the improvements in unlubricated wear performance for all the nitride materials in Figure 3a. With increasing substrate bias, the load support offered by the coating is greater and wear of the coating itself is reduced. However, the corresponding increase in internal stress with bias results in a reduction in wear resistance at high bias voltages due to its adverse effect on coating/substrate adhesion[30].The most wear resistant coatings have zone-T microstructures with low levels of internal stress.

Coating microstructure can also have a more direct effect on the wear behaviour of a coated component. For instance in the sliding wear of coatings with zone-1 microstructures against steel spheres, the open boundaries between the columns can trap metallic debris that has been cut from the sphere by the sharp column tips . As the amount of iron trapped in these intercolumnar spaces increases, adhesion between the sphere and the coating and hence the coating wear rate (at least any adhesive wear component) is also increased. Although the intrinsic adhesion of all the nitride coatings investigated in this study on the steel counterface is low and adhesive wear is not the dominant wear mechanism for denser coatings with the zone-T microstructure, for these low bias coatings adhesive wear is the dominant mechanism as material is pulled out from the coating due to the adhesion of the sphere to the transferred iron that is keyed-in to the microstructure.

The smoother surface of the high bias coatings are thus very important in dictating the wear rate of the coatings in unlubricated sliding and a coating with a dense zone-T structure and low residual stress gives the best wear performance. In the unlubricated test results presented here there is little to choose between the performance of any of the nitride coatings at deposited at -60 V bias which was identified as the optimum level for TiN coatings previously[30].However, in the case of lubricated sliding the shear stresses on the coating/substrate interface produced by the sliding friction are reduced and the microstructural effects at low bias voltages and the stress effects at high bias voltages on the wear performance are much reduced. In this case there is a considerable improvement in wear performance for the zirconium nitride films probably related to their higher hardness.

For corrosion or oxidation protection applications a dense adherent film is paramount as it is essential that the coating completely separates the environment from the substrate material if the optimum performance is to be achieved.In general coatings with the zone-T structure and relatively low levels of residual stress are thus required for corrosion applications in just the same way that they are required in tribological applications.

5 CONCLUSIONS

A knowledge of the microstructure of the coating and how this is affected by the choice of deposition technology and process parameters is essential if the performance of a coated component is to be fully understood. The microstructure/property relationships for a range of SIP nitride coatings are broadly similar. In all cases the transition from the open zone-1 microstructure to a dense zone-T structure leads to an increase in the density and hardness of the coatings. However, the increases in residual stress which lead to reductions in adhesion for titanium nitride films do not have such a great effect on adhesion for the other nitride materials.The wear performance of the coatings depends on both the load support offered by the coating and its adhesion to the substrate.

For SIP nitride coatings the process parameters used for titanium nitride coatings provide a useful starting point for the investigation of new coating materials. However, for other coatings the composition of the sputtering gas is more important for other coating materials where there are a range of possible nitrides and this will also need to be considered if the properties of the coatings are to be optimised for a given application.

ACKNOWLEDGEMENTS

This work was performed as part of the Corporate Research Programme of AEA Technology. The authors would like to thank Jennie Cullen, Peter Fernback and Phil Warrington for performing the coating depositions and wear, scratch and hardness testing measurements.

REFERENCES

1. H.E. Hintermann, Wear, 1984, 100, 381.
2. A. Matthews, Surf. Eng, 1985, 1, 93.
3. D.S. Rickerby and S.J. Bull, Surf. Coat. Technol., 1989, 39/40, 315.
4. J.-E. Sundgren, Thin Solid Films, 1985, 128, 21.
5. D.S. Rickerby and P.J. Burnett, Thin Solid Films, 1988, 157, 195.
6. K.H. Habig, J. Vac. Sci. Technol., 1986, A4, 2832.
7. D.T. Quinto, J. Vac. Sci. Technol., 1988, A6, 2149.
8. S.J. Bull and D.S. Rickerby, Surf. Coat. Technol., in press (1992)
9. R. A. Dugdale, "Proc. Int. Conf. on Ion Plating and Allied Techniques", CEP Consultants, Edinburgh, 1977, p.177.
10. R. A. Dugdale, Thin Solid Films, 1977, 45, 541.
11. J. P. Coad and R. A. Dugdale, "Proc. Int. Conf. on Ion Plating and Allied Techniques", CEP Consultants, Edinburgh, 1979, p.186.
12. D. S. Rickerby and R. B. Newbery, Vacuum, 1988, 38, 161.
13. P. J. Burnett and D. S. Rickerby, Thin Solid Films, 1987, 148, 41.
14. P. J. Burnett and D. S. Rickerby, Thin Solid Films, 1987, 148, 51.
15. S. J. Bull and D. S. Rickerby, Surf. Coat. Technol., 1990, 42, 149.
16. S. J. Bull, D. S. Rickerby, A. Matthews, A. Leyland, A. R. Pace and J. Valli, Surf. Coat. Technol., 1988, 36, 503.
17. S. J. Bull and D. S. Rickerby, Surf. Coat. Technol., 1991, 50, 25.
18. S. J. Bull, D. S. Rickerby and A. Jain, Surf. Coat. Technol., 1990, 41, 269.
19. S. J. Bull and P. R. Chalker, Surf. Coat. Technol., 1992, 52, 117.
20. B. D. Cullity, "Elements of X-ray Diffraction", Addison-Wesley, Reading, MA, 2nd. Edn., 1978, p.447.
21. D. S. Rickerby, B. A. Bellamy and A. M. Jones, Surf. Eng., 1987, 3(2), 138.
22. D. S. Rickerby, A. M. Jones and A. J. Perry, Surf. Coat. Technol., 1988, 36, 631.
23. D. S. Rickerby, B. A. Bellamy and A. M. Jones, Surf. Coat. Technol., 1988, 36, 661.
24. R. J. Jaccodine and W. A. Schlegel, J. Appl. Phys., 1966, 37(6), 2429.
25. G. Stoney, Proc. Roy. Soc. (London), 1909, A82, 172.
26. B.A. Movchan and A.V. Demchishin, Fiz. Met. Metalloved., 1969, 28, 653.
27. J.A. Thornton, Ann. Rev. Mater. Sci., 1977, 7, 219.
28. D.S. Rickerby, J.Vac. Sci. Technol., 1986, A, 2809.
29. P.J. Burnett and D.S. Rickerby, Surf. Eng., 1987, 3, 195.
30 S.J. Bull, D.S. Rickerby, T. Robertson and A. Hendry, Surf. Coat. Technol., 1988, 36, 743.

2.5.5
Transverse Rupture Strength of CVD, PVD, and PVCD Coated WC–Co and WC–TiC–Co Cemented Carbides

S. Zheng, R. Gao, and J. He

DEPARTMENT OF MATERIALS SCIENCE AND ENGINEERING, XIAN JIATONG UNIVERSITY, XIAN, 710049, PEOPLE'S REPUBLIC OF CHINA

1 INTRODUCTION

Vapour deposition techniques, either chemical or physical, by which a few microns of hard metallic or non-metallic compounds can coat a base material, are effective means to enhance wear and thermal resistance, whilst maintaining the toughness of tool materials.

The detrimental effects of vapour deposited (especially high temperature CVD) coatings on the transverse rupture strength (TRS) of cemented carbides have been well documented[1,2]. The deterioration in TRS may result from: improper selection of the base materials, improper design of the coating system or improper deposition process. The present work sought an understanding of how the TRS properties of TiN coated WC-6Co and WC-14TiC-8Co could be altered by the characteristics of the coating (microstructure, thickness, decarbonized zone width and residual stress) and deposition process (CVD, PVD and PCVD).

2 EXPERIMENTAL DETAILS

The two types of cemented carbides used were WC-6Co and WC-14TiC-8Co. The specimens for TRS testing were prepared in the following way: mould pressed and sintered in hydrogen atmosphere and ground to a nominal size of 5x5x30mm. The vapour deposition processes were scheduled as in Table 1.

TRS tests were carried out on an INSTRON-1195 type machine with three-point bending, in which the span distance was 24mm and the loading rate was 0.2mm/min according to the National Standard for the Determination of Transverse Rupture Strength of Hardmetals (GB 3851-83, equivalent to ISO 3327-82). The phase analysis and residual stress measurements in different coating systems were made with a Regaku D/max-RA X-ray Diffractometer. TRS fractography and microstructure analysis were performed using an AMRAY-1000B type scanning electron microscope (SEM).

3 RESULTS AND ANALYSIS

Results from TRS testing of cemented carbides with different coating systems are shown in Figure 1, for comparison the data for the base materials are also plotted.

Evidently PVD-TiN coatings will maintain the original TRS level, but for CVD coatings, TRS declines with an increase of the coating layer numbers, to 66% and 75% in the triple layer deposited on WC-6Co and WC-14TiC-8Co alloys respectively in comparison with the base materials, but recovers a little for seven layer coatings. For PCVD-TiN coatings, TRS results stand intermediate between values for PVD and CVD coatings. In the as-coated states the WC-6Co alloy displays lower values of TRS than the WC-14TiC-8Co alloy, especially in CVD multilayer coatings.

Base Material

The TRS values for bare specimens of WC-6Co and WC-14TiC-8Co alloys are 1560 and 1460 MPa respectively. From the viewpoint of chemical composition, there are two aspects distinguishing these two alloys, WC-14TiC-8Co contains titanium carbide and more cobalt than WC-6Co. In general, the addition of TiC, which is more brittle than WC, into cemented carbides will diminish their TRS values. The more the amount of TiC replacing WC, the more the TRS reduces. On the other hand, when Co which has both a plastic and fracture-resistant matrix phase, is added, it will compensate for the drop in TRS of TiC-containing alloys. In particular, the different contents of titanium carbide and cobalt may result in different TRS values, either in the uncoated or coated situation, for the two alloys.

Coating Microstructure

For CVD coated cemented carbides, the grain size becomes finer with increasing layers of coatings. This is because the grains in the new layer renucleate, but do not extend along the same orientations, on the primary grains of the standing layer. The intersection of the different layers may restrain the growth of the columnar crystals, thus the decrease of the strength and toughness of the coated specimen will not be too severe. In addition, with the increase of CVD coating layer numbers, crystal orientation is changed, a strong (220) orientation appears in the single layer, but the bottom of the triple layer coating is only partly (220) oriented and its top layer is misorientated. For the seven layer coating no preferred orientation has been found in the different layers. PVD-TiN coatings are of very fine grains with strong (111) preferred orientation. PCVD-TiN coatings exhibit fine grains with the usual strong (200) orientation[2].

η Phase and Decarbonized Zone Width (DZW)

During chemical vapour deposition, the following reaction is unavoidable, even if it is processed at a low temperature[3]:

$$TiCl_4 + C + 2H_2 \rightarrow TiC + 4HCl \qquad (1)$$

the carbon needed in the reaction comes from the cemented carbides in the surface layer of the substrate, which leads to the formation of brittle compound carbide η phase (W_3Co_3C) and causes a decrease in the TRS[1,2] and coating substrate adhesion. Such decarbonization of the substrate surface results in a decarbonized zone width at the coating/substrate interface. A test simulating the high temperature deposition process was carried out in which both alloys were heated to 1050^{0}C for 4 hours in a vacuum. It was shown that the fraction of η phase in WC-6Co was 3.6% and 1.5% for WC-14TiC-8Co. Probably, TiC is chemically more stable than WC and the η phase forms with some difficulty. The ease with which WC-6Co decarbonizes, as indicated by the large DZW and the significant decrease in TRS, is well recognized by manufacturers.

DZW, and its relationship to TRS, is also dependent on coating composition, deposition temperature and time.

For instance, the single layer CVD-TiN coating is of medium size as regards DZW and shows an appreciable decrease in TRS; in comparison with the uncoated substrates, the PCVD-TiN coated specimens display a limited DZW and accordingly only a small decrease in TRS. However, for CVD multilayer coatings, the DZW is extensive - because of TiC present in the top of the substrate and the extended dwell time (to achieve a thickness of 6-8 μm) at the high deposition temperature - and consequently the TRS decreases sharply.

Coating Thickness

Several experiments[1,2] demonstrated that the values of TRS of coated cemented carbides decrease as the coating thickness rises, since thermal cracks and other defects increase owing to the difference of thermal expansion coefficients between the substrate and the coating. Moreover, a deep DZW at the coating/substrate interface during the formation of a thick coating layer would aggravate the decrease in TRS.

In the present study, referring to Table 2, for CVD TiC-TiN double layer coated alloys with coating thicknesses of up to 11-12 μm, the decline in TRS is less than that of the CVD triple layer and seven layer coated alloys with thinner coatings of 6-8 μm. It seems that, although all of the CVD coatings were processed in the similar temperature range, the former coating was produced at atmospheric pressure with a high deposition rate of 3-4 μm/h, i.e. deposition time was short; while the latter coatings prepared by low pressure CVD with a low deposition rate of 1-1.5 μm/h, took longer.

Therefore, among the coating factors influencing the TRS of the coated cemented carbides, DZW is predominant. Only when the coating thickness increases with an increase of DZW, does the effect of coating thickness become significant.

Residual Stresses in Coatings

Residual stress measurements of the different coating systems are listed in Table 2. Residual stress in the coating layer will, to some extent, affect the transverse rupture properties of the coated cemented carbides. For

PVD-TiN coatings, high levels of compressive residual stresses will keep the TRS of the substrate from decreasing. For PCVD-TiN coatings, low levels of compressive residual stress are associated with a small decrease in TRS. For CVD coated alloys, the residual stresses change from tensile in the single layer coatings to compressive in the multilayer coatings, and prevent the TRS from further decreasing.

Fractography

The fracture characteristics for TRS tests of the coated cemented carbides can be analysed from the rupture surface by means of SEM fractography. For CVD coated alloys, the fracture origin was located at the coating/-substrate interface (see Figure 2). It seems that η phase plays an important rôle in promoting the transverse rupture. PVD-TiN coated specimens showed a sub-surface cracking source of fracture similar to that of the base materials and a constant TRS value was revealed. For PCVD-TiN, the coated specimens behaved like that of CVD coated ones, except that the former had even smaller DZW and consequently a little higher TRS values.

4 DISCUSSION

Selection of Base Material

With reference to Figure 1, the cobalt content in cemented carbides to be coated should be a little higher, e.g. 9-12%, since Co plays an important rôle in retarding the decrease of TRS. Furthermore, additions of TiC, TaC and/or NbC to substitute partly for WC in WC-Co base alloys will, to some extent, reduce the sensitivity to decarbonization of cemented carbides during vapour deposition at high temperature and maintain high TRS and cutter-edge strength.

Modification of Deposition Process

PCVD, in combining advantages of conventional CVD and newly-developed PVD, offers great potential to cemented carbides in that, the deposition temperature can be adjusted to a wide range 100-1000^{0}C, and the chemical composition of coatings can be changed easily giving good coating/substrate adhesion[2]. Here, an improved PCVD approach is proposed as follows: first, to prepare the TiC bottom layer at higher temperature (e.g. 800-900^{0}C for an appropriate time to obtain better coating/substrate adhesion as well as a thin η phase layer; secondly, to produce TiCN···TiN layers at lower temperature(e.g. 400-600^{0}C) in order to avoid forming a thick η phase and achieve a coating of high quality.

5 CONCLUSIONS

CVD coated cemented carbides show that, with increasing number of layers of coatings, the grain size becomes finer and the degrees of preferred orientation are reduced gradually; the residual stress transforms to compression in the multilayer coatings from tensile in single-layer ones. The width of decarbonized zone at the coating/substrate interface is the dominant factor in determining values of TRS.

PVD-TiN coated cemented carbides consist of very fine grains with strong (111) preferred orientation in the coatings, and nearly maintain the original level of TRS since high compressive residual stresses are involved.

PCVD-TiN coated cemented carbides exhibit strong (200) preferred orientation in the coatings. An intermediate TRS value, somewhere in between those of CVD and PVD results from the small DZW at the coating/substrate interface.

In as-coated states, WC-6Co alloy decreases in TRS less than WC-14TiC-8Co alloy, especially when multilayer coatings with TiC deposited as the first layer are involved. This is attributed to the high sensitivity of decarbonization and a thick η phase layer formed at high temperature for WC-6Co alloy.

High TRS and better coating/substrate adhesion in the vapour-deposited cemented carbides would be achieved by PCVD coating TiC as the first layer at higher temperature and then -TiCN···TiN at lower temperature, since the decarbonized zone is not so great and the residual stress is compressive.

Table 1 Specimen type and vapour deposition process

specimen No	coating system	deposition temp., ℃	deposition rate, μm/h
PVD	TiN	500	~1.0
PCVD	TiN	500—600	~3.0
CVD—1	TiN	950	~1.0
CVD—2	TiC—TiN	950—1050	~3.0
CVD—3	TiC—TiCN—TiN	950—1050	~1.0
CVD—7	TiC—TiCN···TiN	950—1050	~1.0

Table 2 Measurements of coating thickness and residual stress in coatings

		PVD	PCVD	CVD—1	CVD—2	CVD—3	CVD—7
coating thickness, μm		1.6	3.0	4.4	11.5	8.2	6.3
residual stress, GPa	WC—6Co	—4.77	—0.89	+0.98	—0.95	—0.33	—
	WC—14TiC—8Co	—5.33	—0.69	+0.70	—	—0.67	—

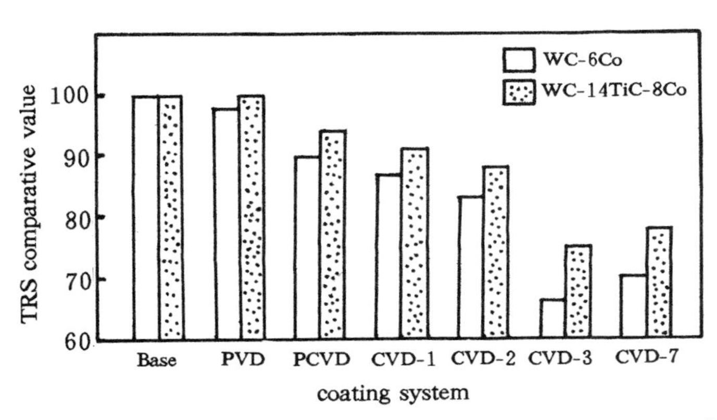

Figure 1 Schematic diagram of TRS test results

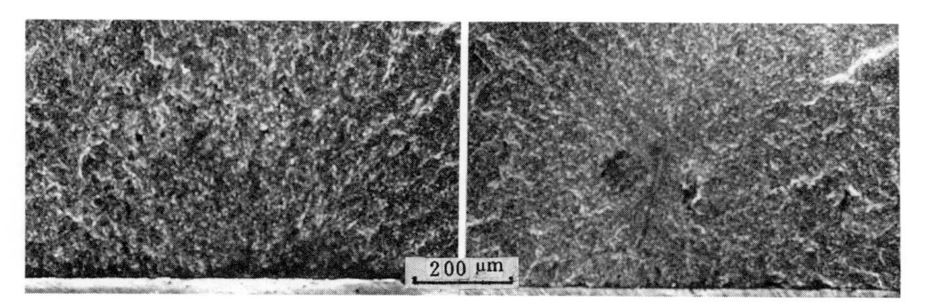

Figure 2 TRS macrofractograph of CVD-trilayer coated WC-6Co

Figure 3 TRS macrofractograph of PVD-TiN coated WC-6Co

ACKNOWLEDEMENTS

The authors would like to thank Ms Cheng Nanru for her help in X-ray analysis, and the Laboratory of Solid Lubrication in Lanzhou Institute of Chemical Physics, Chinese Academy of Sciences for supporting this work.

REFERENCES

1. Liu Guochun, Cemented Carbides, 1989, 3, 4-7.
2. Liu Guochun and Bian Hengzheng, Cemented Carbides, 1991, 2, 20-23.
3. MY Al-Jaroudi, HTG Hentzell and SE Hornstrom, Thin Solid Films, 1989, 182, 153-166.
4. PC Jindal, DT Quinto and GJ Wolfe, Thin Solid Films, 1987, 154, 361-375.

Conference Photographs

The following plates show speakers, guests and delegates at the 3rd International Conference on Advances in Coatings and Surface Engineering for Corrosion and Wear Resistance, and the 1st European Workshop on Surface Engineering Technologies and Applications for SMEs. Both events were organized by the Surface Engineering Research Group of the University of Northumbria and took place between 11-15th May 1992 at Newcastle upon Tyne, UK.

City of
Newcastle upon Tyne

TECVAC

Sheffield
City Polytechnic

Contributor Index

This is a combined index for all three volumes. The volume number is given in roman numerals, followed by the page number.

Subject Index

This is a combined index for all three volumes. The volume number is given in roman numerals, followed by the page number.